$27.50

Landsat 1 photograph of the Salt Lake City area taken October 13, 1973. Snow is visible in the mountains, and the downtown area shows up as gray to the right of the lowest lobe of the larger lake (Great Salt Lake). The rectangular patterns of cultivated fields appear faintly in scattered areas in the center and larger agricultural areas near the mountains. This false-color image shows healthy vegetation as red. (U.S. Geological Survey/EROS Data Center)

Viewing the Earth

Inside Technology

Wiebe E. Bijker, W. Bernard Carlson, and Trevor Pinch, editors

Artificial Experts: Social Knowledge and Intelligent Machines
H. M. Collins, 1990

Viewing the Earth: The Social Construction of the Landsat Satellite System
Pamela E. Mack, 1990

Inventing Accuracy: A Historical Sociology of Nuclear Missile Guidance
Donald MacKenzie, 1990

Viewing the Earth
The Social Construction of the Landsat Satellite System

Pamela E. Mack

The MIT Press
Cambridge, Massachusetts
London, England

353.00856
M15v

© 1990 Massachusetts Institute of Technology

All rights reserved. No part of this book may be reproduced in any form by any electronic or mechanical means (including photocopying, recording, or information storage and retrieval) without permission in writing from the publisher.

This book was set in New Baskerville by The MIT Press and printed and bound in the United States of America.

Library of Congress Cataloging-in-Publication Data

Mack, Pamela Etter.
    Viewing the earth: the social construction of the Landsat satellite system / Pamela E. Mack.
    p. cm.—(Inside technology)
    Includes bibliographical references.
    ISBN 0-262-13259-1
    1. Landsat satellites—History. 2. Astronautics and state—United States. I. Title. II. Series.
TL796.5.U6L375   1990
353.0085'6—dc20
                                                                        90-32074
                                                                         CIP

Contents

Preface ix

I
CONTEXT

1
Technological Change in the Government Context   3

2
The Space Agency and Applications Satellites   15

3
Sources of Interest in Earth Resources Satellites   31

II
DEVELOPMENT

4
NASA Establishes a Program   45

5
User Agencies Seek a Role   56

6
Selecting the Sensors   66

7
The Battle for Funding   80

8
The Design Phase: Organizational and Technological Choices   94

9
Data Processing: The Technical Challenge   107

III
APPLICATION

10
The Transition to Developing Practical Applications   121

11
The Department of the Interior's EROS Data Center   132

12
Agricultural Applications of Landsat Data   146

13
Technology Transfer and State and Local Governments   159

14
Private Industry Users and the Early Debate over Privatization   *171*

15
International Issues for Earth Resources Satellites   *183*

16
Early Steps toward an Operational System   *196*

Epilogue   *208*

Notes   *212*
Notes on Sources   *248*
Index   *259*

Preface

This study of Landsat grew out of curiosity about practical applications of the space program. My initial question was, How can the government most effectively sponsor technological innovation? It has become clear that the government needs to sponsor development of certain advanced technologies with broad benefits when private industry cannot afford such development. In NASA, at least, the government has sometimes been an effective sponsor of technological advance. More problematic for the space agency has been the process of bringing newly developed technology into practical use. Commentators have often viewed ineffective use of technology simply as a result the inherent inefficiency of government, or as a result of specific technical faults or organizational problems. Rather than using just one of these factors to analyze effectiveness, this study explains it on the basis of the conflict among interest groups over not only control of technological innovation but also the definition of the new technology.

This study focuses on the role of the potential users of a technology during its research and development. The relationship between research and development and use is particularly complex in the government context because the question of whether an innovation will sell is not as central as in a business context. My initial question was whether government-sponsored technological advance could be made more beneficial by tying it more closely to the needs of the ultimate users. Again, however, the issue turned out to be more complex. Problems arose not necessarily because potential user organizations did not have a voice but because different interest groups had such different understandings and expectations of the project that they could not reach an effective compromise among their various needs.

Scope of the Study

This study examines the Landsat project, which developed and tested satellites to monitor earth resources, from its origins through the launching and use of the first few satellites. The process by which Landsat later actually became operational and was turned over to private industry is an important one, but the outcome of that process is still so uncertain that it would be foolhardy to try to write its history. Even as of 1990 it was not clear that the U.S. earth resources satellite program would ultimately be successful, although the turning over of Landsat to private industry and the early successes of the French SPOT satellite (whose development occured after the period covered in this study) provided strong evidence of the potential for success of an operational earth resources satellite system. My argument is that the development of a technology cannot be planned or understood in isolation from its eventual use and the interests of the users, but my subject here is the role of the users in the experimental stages of the project, not the transition from an experimental program to a fully operational one.

As a historian of technology, my central interest is not in how policy is made but in how policy affects how technology is developed and used. Such a study is inherently interdisciplinary, but it is impossible to treat all aspects to the same depth. My analysis concentrates on NASA and the agencies that used Landsat, not on how decisions were made by the president, the Bureau of the Budget, and Congress. Obviously those other institutions play a role in my story, but their inner workings are not my focus.

The Role of the Individual

Although my interest as a historian lies in finding patterns in the complexities of a particular case study, the nature of recent government-sponsored technology limits the attention to individuals that is traditionally a trademark of history. Sometimes an individual leaves an indelible impression on a project (for example, Admiral Hyman Rickover and the nuclear submarine), but many large technological projects are shaped more by the forces of bureaucracy than by the forces of personality. It is important not to treat NASA as an organizational black box, but the conflicting interests within the agency are more often those of different divisions and programs, not those of different individuals.

In addition, in the recent history of government agencies it is often difficult to conclusively attribute decisions and opinions to particular individuals. In many cases the minutes of key meetings are unavailable or lacking in detail. Even letters often do not reveal individuals; frequently the writing of a letter is assigned to someone other than the person who signs it. Interviews, of course, provide a personal view, but one that is often colored by hindsight. Therefore, the narrative that follows, while highlighting a few individuals, focuses on the role of interest groups and bureaucratic forces in shaping technology.

This, of course, risks treating an agency made up of people with different opinions as if it speaks with a single voice. To avoid this problem I have attempted to be consistent in the use of three types of voices. Most often I have found it necessary to describe opinions as those of NASA leaders or NASA managers, implying that those opinions were the most common and influential ones, although there may have been dissenters. Where relevant and possible, I specify an opinion as that commonly held in a particular department or center within the agency. In some cases, however, I use NASA as the subject of a statement, implying that the opinion in that statement was the official policy of the agency. It should be noted that particularly in budgetary matters and testimony before Congress, the official policy of the agency may have been set by the executive office of the president and may not have reflected the opinions of NASA management.

Acknowledgments

In the twelve years since I started my study of Landsat, I have accumulated more intellectual and personal debts than I can properly acknowledge. The History Office at NASA Headquarters in Washington, D.C., provided summer internships for three summers, and the National Air and Space Museum of the Smithsonian Institution provided a Daniel and Florence Guggenheim Fellowship for one year. NASA's Office of Applications, Virginia Polytechnic Institute and State University, and Clemson University provided travel money at various times. My research was made possible by agencies and individuals who helped me locate and make sense of the documentation. Lee Saegesser at the NASA History Office was my most important guide. I also had the help of staff at the Johnson Space Center History Office, the National

Archives, and the Washington National Records Center. Scientists, engineers, and managers at NASA, the Department of the Interior, and the Department of Agriculture shared their views in interviews and let me go through their file cabinets and stored records that had not yet been retired to the federal records centers. A number of participants read parts of the manuscript in various stages and provided valuable corrections and comments.

Alex Roland and Thomas P. Hughes played important roles in guiding and nurturing this study. I am also particularly grateful for substantive advice from John M. Logsdon, Edward C. Ezell, David H. DeVorkin, Michael M. Sokal, W. Bernard Carlson, John W. Mauer, and unnamed referees. I thank the whole community of space historians for providing me with both a lively intellectual context and a gang of friends. Family and friends have supported me in many ways during the course of this study; I hope they know my gratitude more clearly than I can express it here. I owe special thanks to those who helped me with the final stages of preparation of this book when I was slowed down by pregnancy: John W. Mauer, Lorna C. Mack, and Nancy A. Hardesty.

I
Context

1
Technological Change in the Government Context

In 1972 the National Aeronautics and Space Administration (NASA) launched a satellite to take unclassified pictures of the earth for resource management. Scientists were excited by the potential of the new satellite: Hydrologists used the data to measure the extent of snow cover, agricultural scientists looked for diseased crops, geologists developed new ways of finding likely sites for oil exploration, and geographers discovered new coral reefs and islands. Orders for satellite data came from government agencies, universities, industry, foreign governments, and high school students who wanted a satellite picture of their home town. Enthusiasts expected that this first Landsat satellite would quickly lead to either a routine government earth resources satellite service, on the model of weather satellites, or a successful commercial use of space, on the model of communications satellites. Instead, as of 1990, the U.S. earth resources satellites program is limping through the early stages of commercialization, whereas a French private-public partnership is selling data from earth resources satellites more advanced than any civilian satellite launched by the United States. The reasons that the U.S. earth resources satellite program has so disappointed its backers can be understood only by understanding how technological change is negotiated among organizations and groups of individuals who may have very different goals.

Historians of technology deal with the process by which people develop and apply new technology. Obviously, technological innovation varies depending on the individuals, groups, and organizations involved in the development of a particular technology. Most simply, different groups of engineers with different skills would approach the same problem in different ways. Scholars have recently begun to examine in a broader way how technological innovation is shaped by its participants. In particular, the influence of various interest groups on technological innovation involves not only different technical approaches but also different social goals and ways of understanding the hardware itself. Scholars taking this approach have argued that the development of new technology involves a process of social construction, by which a community of interested individuals and organizations negotiates a definition of the character and goals of the new technology.[1]

Even in this context, historians have concentrated on the developers of new technology and have given little attention to the role of the eventual users of a technology in shaping innovation by their needs, their interests in particular technologies, and their predictions of how the technology will be used.[2] Yet if the development of technology involves the negotiation not only of how best to achieve particular goals but also of what those goals should be, then the potential users play a key role in the process (it is possible for the developers of a technology to exclude the potential users from the process of defining that technology, but doing so is likely to have clear consequences as the technology matures). The case of the Landsat satellite project provides insight into the role of users in defining a new technology and into the process of technological change in the government context.

The Case Study

Earth resources satellites provided new information about our world by collecting data in the form of sophisticated images of the surface of the earth. Interest in such a satellite first arose in the early 1960s among scientists working with data from classified reconnaissance satellites. NASA developed a project originally called ERTS, for Earth Resources Technology Satellite, and then changed the name to Landsat in 1975. (NASA managers joked that they had to change the name because when they said "erts" the most common response was "gesundheit.") The development of Landsat quickly became tangled in conflicts based on budgetary issues, security concerns, divergence of interest between the developers of the technology and the potential users, and bureaucratic competition. These problems shaped satellite design and also slowed the project. The end result was a fairly small (941 kilogram) satellite, first launched in 1972, carrying two sensors that produced images of the earth by different techniques. Although the new satellite provided data of coarser resolution than that provided by Department of

Defense reconnaissance satellites, Landsat data were useful for studies in geology, agriculture, and resource management fields, such as land use.

The development of Landsat involved a range of controversies illuminating the role of users in technological change. Because the development and the use of earth resources satellites were split up between NASA and a variety of government agencies and other institutions, interagency politics served to mediate between the interests of the research and development team and a wide range of user interests.[3] Space agency managers needed outside support to get authorization and funding for an earth resources satellite. Bruno Latour has analyzed the dangers of such piggybacking of interests: "Sometimes you have to overcome the indifference of the other groups (they refuse to believe you and to lend you their forces), and sometimes you have to restrain their sudden enthusiasm."[4] NASA leaders faced both problems with potential users for Landsat data.

Even in the early stages of the Landsat project, controversy revolved around disagreements over what voice the users of the satellite data should have in the development of the project and at what point it should be turned over to the users. Starting early in the project, NASA provided funding to research teams at potential user agencies, such as the U.S. Geological Survey, to develop applications for Landsat data. Some of these agencies wanted a larger role in Landsat, but NASA tended to want to keep control of the project and the design and limit the users to the development of applications. For example, one early debate centered on whether research and development would be conducted to produce a satellite best suited to the immediate needs of the users or one that the researchers believed would have the most potential.

Even after the successful launch of the first satellite, the project faced many difficulties. The satellite functioned well and scientists found that the data yielded much valuable information, yet potential users only slowly developed systems to use Landsat data effectively. Although some potential users fought for a larger role in the project, many showed much less interest in Landsat data than predicted. The resulting low demand for the data made it difficult to obtain funding for an operational system. Space agency managers found themselves under pressure to promote use of the data, a task for which they were ill prepared. The problems of encouraging use of Landsat data grew not only out of space agency managers' lack of understanding of technology transfer but also out of divergent approaches to using the data, particularly between the space agency and many potential user organizations.

Conflict also grew up over who would manage an operational satellite program. Although the space agency had the responsibility for research and development of space applications, NASA usually turned over operational programs (those ready for routine use) to the agency concerned with the related application (e.g., the Weather Bureau took over weather satellites). However, the process was not so simple in the case of Landsat, which President Carter decided to turn over first to the National Oceanic and Atmospheric Administration and then to private industry. Planning for an operational system also involved larger questions of appropriate uses of technology developed by the federal government: Should data from civilian remote sensing satellites be sold at a profit or provided as a public service?

Large-scale, government-funded technology provides an ideal environment for studying the role of users in technological change. Organizations look after their own interests, so new technology is shaped by the process of negotiation among the interested organizations. Unlike the users of consumer technology, the users of Landsat belonged to institutions that communicated according to set bureaucratic patterns and generated plentiful documentation. In addition, potential users played a particularly important role in Landsat because the technology involved not radically new hardware but rather fairly straightforward combinations of existing equipment and ideas, so that its uses could be more easily anticipated. Interagency politics is not necessarily destructive, but in the case of Landsat a series of particularly bitter fights developed out of the pressures caused by general political opposition to the space program in the late 1960s and 1970s. Yet it is by no means clear that Landsat could have been handled better by private industry; NASA had the opportunity to develop the Landsat program to bring the most benefits rather than the most profit.

Historians of technology have tended to look at technology in private industry and at the process of invention; this study concentrates on government technology and on the processes of development and implementation of new technology. Certainly Landsat showed the effect of bureaucratic politics on technological change, but that effect requires more detailed analysis. The government context raises many new issues, including the interaction of government agencies, the tendency of research and development organizations to pursue the most exciting technology rather than the most practical uses, and the difficulties of building a political constituency for technological innovation. Because Landsat was so controversial, its history provides a particularly clear example of how interest groups mold a new technology in a government context. The remainder of this chapter will first examine the process of technological change and then look at how this change is structured in the U.S. government in general and NASA in particular.

Technological Change

Two approaches to technological change will help make sense of the influences shaping the development of Landsat. First, we can break down technological change into phases to provide a set of benchmarks, even if technological change often does not involve a simple linear progression between phases. Second, we need to develop a language for describing who shapes technological change and how.

The process of technological change has often been divided into invention or research, development, and innovation and/or diffusion. Many historians of technology have examined the invention or research phase, and Usher, Schmookler, Gilfillan, and others have written on the theory of invention.[5] T. P. Hughes has built on such work to develop a systematic understanding of the process of technological change, arguing for the importance of examining not just individual inventions but whole technological systems and the economic, political and cultural factors that form an integral part of those systems.[6] He has also called for historians of technology to give more attention to the development phase, when a concept is turned into a usable technology.[7] Looking at the later phases of technological change from another perspective, business historians have analyzed the process of making a new technology salable and mass-producible.[8] Rural sociologists have led the examination of the process of diffusion of new technology into widespread use.[9]

For the Landsat case, technological change can roughly be divided into four stages: researching and developing the satellite and also equipment on the ground used to process data coming from the satellite, testing the system and developing ways of applying the new data, persuading people to use the data and develop new applications, and creating a truly operational permanent system. Obviously these phases involve not just development of hardware but also development of organizations, funding, attitudes, scientific knowledge, and interconnections with other systems.[10]

Each phase of technological change bears a different relationship to the question of how the technology will ultimately be used. Since most technologies originate because someone believes they will be useful or marketable, use is not a factor that enters only in the later stages of the process. Yet, at least in the case of advanced technology developed by research and development teams, the ultimate use does not shape the technology so exactly that the process of integrating the new technology becomes trivial. Rather, the interaction between the push of new technical ideas and the pull of existing needs is a complex process that determines whether a new technology will bring the benefits its promoters predict. A new technology that simply responds to recognized needs is likely to be only a small advance, because users tend to be committed to existing systems and reluctant to take risks. On the other hand, a new technology developed because it is technically exciting rather than because it meets a need is likely to be an expensive failure when it reaches the stage of diffusion into widespread use.[11] Clearly, on a simple level, a successful technology requires good communication among the engineers, scientists, and managers creating the new technology and the ultimate users throughout the entire process of technological change.[12]

However, good communication requires that the parties speak the same language, and there another series of issues arises. Recent studies of technological change have suggested that different groups involved in innovation may have radically different ways of looking at a technology, not just different requirements for a specific innovation.[13] More generally, a new idea becomes scientific or engineering knowledge only by being accepted by enough influential scientists and engineers.[14] The new technology is socially constructed because the communities interested in that new idea shape it to meet their own interests. This approach to understanding innovation has been well developed for the study of science but is fairly new in the study of technology.[15] At its strongest it suggests that an artifact cannot be treated as an objective fact, since it will have different uses and even receive different evaluations of its success or failure from different social groups.[16]

It is not necessary here to argue about the objective reality of technological artifacts. What is useful for the purposes of this study is the insight that influences on technological change must be seen not just in terms of specific technical requirements but in terms of a competition between interest groups over different conceptions of the meaning, use, and approach to a new technology. This study examines how different interest groups define technology for their own purposes and negotiate between conflicting definitions. In this context the verb define means not just setting specifications for a new piece of hardware but rather the process by which a particular group develops its own understanding of what that hardware or technological system can and cannot do and how it should be used.

Government Process

Before looking at alternative definitions of earth resources satellites, it is necessary to understand better the interaction of the government agencies that served as the main interest groups for Landsat. In one sense politics obviously shapes government-sponsored technology, but it is important to ask what kinds of politics have what influences. The word *politics* often includes not only electoral politics but also the relationships among branches and agencies of the government. Using that broad definition, we can list many types of politics that affect government projects. It is crucial to recognize that the government context is far from monolithic; it can be viewed on a variety of levels, from philosophies of government to management policies within an agency. Examinations of the space program by political scientists, such as John Logsdon's study of the origins of the Apollo program, have tended to focus on policy-making at the presidential level.[17] Historians of NASA projects have often concentrated on project management and the process of research and development itself.[18] Clearly, technological change is influenced by many levels of the organizational hierarchy in which it takes place. The Landsat case supplies illustrations of many of kinds of political influences, but it provides a particularly striking example of the importance of bureaucratic politics—a pattern of relationships among government agencies centered on issues of competition and control.

Looking broadly at the government context reveals the influence of competing philosophies of government. One nagging controversy in the Landsat project was whether the whole thing should properly be a government enterprise or whether it should be left to private industry. Ideally, the government sponsored technological development for the public good when the benefits were too diffuse or too far in the future for private industry to be willing to sponsor it.[19] Opinions about what fell within this definition depended on attitudes toward the relative roles of government and private industry.[20]

The highest stage of the government hierarchy played a much smaller role in Landsat than in more expensive and popular space projects. Landsat cost only about $32 million through the launch of the first satellite, so it was too small to get much attention from the highest levels of government. What little authority was exerted over Landsat from above NASA came mostly from the president through the Office of Management and Budget, which negotiated with federal agencies their share of the budget the president proposed to Congress. The Office of Management and Budget opposed Landsat both because of a presidential policy seeking to cut government spending and because of specific opposition at the budget bureau to earth resources satellites (see chapter 7). Landsat received occasional support from the presidential level when NASA leaders appealed to Presidents Nixon and Carter to prevent devastating budget cuts. But each president who gave the project support also expected it to serve his own policy interests, whether they were improving international cooperation or diminishing the size of big government. Congress, particularly the House Subcommittee on Space Science and Applications under the leadership of Congressman Karth, gave the project more consistent support. However, Congress could not oversee the project in as much detail as could the Office of Management and Budget. Congress occasionally reversed a damaging budget cut, but NASA's top policymakers could not appeal directly for such a cut because they had to be loyal to the administration that had appointed them. Thus, the President and Congress affected but did not determine the policies of the agencies interested in Landsat.

In the case of a small project like NASA, it is common for government agencies to make some of their own policy rather than simply carry out the policy of the president. Political scientist W. Henry Lambright has shown the importance of bureaucratic momentum in modern American technology policy. Projects that are rejected by the president and the policymakers he or she appoints stay alive in the ranks of career civil service managers until the political climate becomes more favorable.[21] For example, project Apollo was rejected by President Eisenhower as too expensive, but it was ready and waiting when President Kennedy asked NASA for a feasible space spectacular. Similarly, the Landsat project survived even though it never had substantial support from elected officials and their appointees.

Since Landsat got little attention from political decision makers, decisions had to be thrashed out among the interested agencies, with NASA taking a dominant role as lead agency for the project. The Landsat project was shaped by two kinds of interagency politics. One of the functions of government bureaucracy is to make decisions on grounds of the good of the nation rather than just profit or technological feasibility. The process of interagency politics at its best is a process for negotiating decisions among competing priorities for the use of scarce resources. Because of the many interests involved, the Landsat project required a large amount of this kind of decisionmaking.

But Landsat was also influenced by another kind of interagency politics. Government agencies seek to increase their own importance by increasing their budgets and the number of projects they control (they may have to do this simply to obtain the influence necessary to achieve their broader missions), and a project can be hurt or helped by this competition for power for reasons independent of the merit of the project.[22] In an interagency project, each agency is monitored by its branch of the Office of Management and Budget and its congressional committee to determine how well that agency uses its funds to meet its objectives. Few mechanisms exist for rewarding agencies that work together effectively toward larger goals. In the Landsat case NASA had to deal with the different interests of the Departments of Defense, the Interior, and Agriculture and the Office of Management and Budget. Agencies had interests both in whether stress was put on the experimental or operational nature of the project and in specific technologies. For example, the Department of Defense did not want reconnaissance satellite technology revealed by civilian use; the Department of Agriculture wanted the satellite to carry sensors that would be most useful for agricultural research and survey. Interagency politics caused special problems for Landsat because NASA managers had little experience in working cooperatively with other agencies on practical projects.

In the Landsat project, NASA had to deal not only with other government agencies but also with a variety of organizations that the space agency hoped would eventually make practical use of Landsat data. NASA had little experience in dealing with constituencies and clients other than politicians, scientists, and the aerospace industry. NASA sponsored a variety of programs to encourage the use of Landsat data, but agency leaders were disappointed by the results. The interactions of technology and context were particularly clear in the case of NASA's attempt to encourage the use of Landsat data by state and local governments (see chapter 13). Such users could give NASA valuable political support, but they were hard to convince to use Landsat data because they were particularly inexperienced with and suspicious of advanced technology. The importance of skillful technology transfer and early involvement of the users is one of the most important lessons to be learned from Landsat.

NASA Structure

NASA managers followed a characteristic pattern in the way they sought to shape technological innovation for Landsat, a pattern that was in some ways specific to NASA and in other ways typical of government research and development organizations. Most government research and development is sponsored by agencies like the Department of Agriculture, which also have operational responsibilities for regulation and management. The National Science Foundation and NASA are exceptions, agencies whose primary mission is research and development. Since the nineteenth century, Congress and scientists have repeatedly rejected the idea that all government-sponsored research should be administered by a cabinet-level Department of Science. Opponents of the idea of a Department of Science have argued that in an agency without a direct practical mission, research could not easily be justified by or responsive to its ultimate practical applications.[23] Although concentrating expertise on spacecraft in one organization could certainly be justified, in the Landsat case NASA suffered from exactly the problems predicted for a Department of Science. While research and development groups always tend to be more interested in advanced technology than in practical needs, a research group in a separate agency from the ultimate users is likely to have particular difficulty keeping the concerns of the users in mind. In addition, NASA suffered from funding difficulties because its budget could rarely be tied directly to national needs.

Both the strength and the weakness of a research and development agency like NASA lie in its freedom to be bold in the research and development phases of technological change. As a government agency, NASA could develop Landsat as a project that would bring broad benefits from improved resource management even though it was clearly not going to be profitable quickly. NASA could design a satellite valuable to a wide range of users rather than having operating agencies develop different satellites for slightly different uses. Because NASA was an independent research and development agency, NASA engineers were free to investigate the potential of various technologies before worrying about the requirements of the users. The space agency had an interest in developing the best possible system rather than throwing something together to meet user demands that might reflect only current expectations rather than the true potential of the technology. At its best, this approach can bring unexpected benefits from technological breakthroughs. At its worst, an independent research and development organization can end up pursuing technological sophistication for its own sake when a simpler, less expensive, system would better meet the needs of the users.

The conflicting interests of technology developers and users involved not only definition of the satellite but also definition of the transition from experimental to operational earth resources satellite programs. Research groups and users not only had different priorities but also had different concepts of the process of technological change. Since NASA had responsibility for research and development but turned operational programs over to the appropriate user, the space agency had an interest in prolonging the experimental phase of a project as long as possible, whereas the agencies that would take a project over had an interest in defining the project as operational as soon as possible. But this was not the only use to which definitions of experimental and operational were put. The Office of Management and Budget used a narrow definition of an experimental project to constrain Landsat by requiring that the system built be adequate only for testing the concept, not for routine use. Finally, the definition of the project as experimental for more than ten years scared away potential operational users and reduced the pressure to bring it into effective use. The definition of the phases of technological change thus became a tool of the conflicting interests of the organizations involved.

NASA's situation as a research and development agency also had a particular influence on the process of developing applications and cultivating users for Landsat data. The space agency had little experience with dealing with users outside the agency. NASA managers learned the hard way that publicizing a new technology and funding scientific research to show its potential were not sufficient to convince organizations to adopt it for everyday use. In addition to developing applications, NASA had to learn to work with an assortment of resource management organizations in persuading people to use the data from the satellite.

NASA had little experience with selling new technology, and it showed. It is often assumed that this is an area where private industry has an advantage. Companies have by necessity learned how to put forward an innovation to suit the users and how to persuade organizations to try risky new technology. However, parts of the government have been able to develop impressive skill in encouraging innovation, most notably the extension programs of the Department of Agriculture. Therefore, the problem lies not in government agencies per se but in NASA's role as a research and development agency without a clearly developed relationship with a constituency.

The interaction of technological change and the government context depends on historical circumstances, so the Landsat case study can only raise questions, not prove a model. The Landsat case suggests the importance of examining the influence of the needs of the users on the design of the technology. The question of NASA's responsiveness to the need of potential users can be teased apart into two partially separate stories: the designing of the satellite, and the process of learning how to utilize the results. The next two chapters set up the context in which the Landsat project evolved. Section II traces the development of the satellite technology, and section III focuses on the development of uses for the satellite data. The history of the Landsat project suggests that development of both technology and applications can best be understood as a conflict between interest groups over different definitions of what the technology will do for the users.

2

The Space Agency and Applications Satellites

The roles of the interest groups that shaped Landsat developed in the context of the broader history of the space agency. To understand the history of Landsat it is necessary to evaluate the stage set by the early history of NASA and to compare Landsat with other applications satellite programs. The historical circumstances of the origins of the space program resulted in certain priorities for the agency as a whole and also created influential tensions between NASA and the Department of Defense. More specifically, NASA's other practical applications satellite projects raised questions and set precedents that had a strong influence on Landsat. A comparison of weather, communications, and earth resources satellites shows that Landsat was not the only project to suffer from controversies about how the practical use of space technology should be managed and what voice the ultimate users should have in development. The comparison of Landsat with other applications programs suggests a common pattern: NASA has generally been very successful as a research and development agency, but the agency has shown less skill in working with the ultimate practical users of space technology.

NASA History

The space agency grew out of U.S. reaction to the launch of the first artificial satellite, *Sputnik*, by the USSR on October 4, 1957. From its beginnings, the new agency was shaped by conflicting visions of the purpose of space exploration held by the president, the military services, and the scientific community. The end result of the negotiation among these and other interest groups to define the goals of the space program was that the exploration of space for national prestige received the highest priority, with science taking second priority and practical benefits third. Military uses of space also received high priority, but the creation of separate military and civilian programs usually meant that they were not competing for the same resources.

In July 1958 President Eisenhower established NASA by signing the National Aeronautics and Space Act. Fearful that the cold war between the United States and the USSR would expand into space, the president instructed the authors of the space act to declare that it was "the policy of the United States that activities in space should be devoted to peaceful purposes for the benefit of all mankind." A civilian agency would manage U.S. aeronautical and space activities except those "peculiar to or primarily associated with the development of weapons systems, military operations, or the defense of the United States."[1] Eisenhower was reluctant to approve expensive space projects that could not be justified on military or scientific grounds, but political pressure in reaction to *Sputnik* shaped the space program to give higher priority to national prestige as a goal than Eisenhower wished.[2]

The scientific community, through the President's Science Advisory Committee, played a major role in shaping the legislation that created NASA but had less influence than many scientists had hoped over the resulting agency. Scientists hoped that NASA would be a new scientific research organization in a most promising field, but from the start the leaders of the agency resisted control by purely scientific interests.[3] The new agency needed support, and scientists carried relatively little clout in persuading Congress to increase funding for the space program. This resulted in resentment by many scientists over the fact that the space agency gave higher priority to national prestige than to science.[4]

The military services already had their own space programs before the creation of NASA, so the decision to establish a civilian space agency required a troublesome differentiation between military and civilian space research. According to the space act, all projects and proposals for hardware without a clear military justification should have become the responsibility of the new agency. Predictably, the air force and the army resisted turning over their many plans for spacecraft. Of those programs they eventually relinquished, the services most resented the loss of their largest booster program (Saturn) and their role in developing spacecraft to carry people. The Department of Defense did retain authority over reconnaissance satellite programs—an obviously military enterprise—as well as over an assortment of missile programs for use as weaponry. The Pentagon jealously guarded these remaining programs, creating tensions that greatly affected Landsat.[5]

The young space agency was vitalized by the high national priority given to the space program, especially after John F. Kennedy became president in 1961. Four months into his term, President Kennedy, for reasons primarily of national prestige, committed NASA to landing people on the moon and returning them safely to earth before the end of the decade.[6] This was an ambitious goal for an organization that had yet to put its first astronaut into orbit. Within two and one half years of its creation, then, the new agency had gained coherence around a central challenge with strong public support. That unity and that public support lasted until the landing of people on the moon in 1969.[7]

In the first three years of the new agency, NASA leaders had not only to work out the agency's mission and its relationship with other agencies and potential constituencies but also to develop an effective organization. NASA started out as an amalgam of projects and people transferred from the military, the International Geophysical Year organization and its various projects, and the National Advisory Committee for Aeronautics (NACA). When NACA became the core of the new space agency in October 1958, most of the staff at NACA headquarters in Washington, D.C., and its three research laboratories in Virginia, California, and Ohio became part of NASA.[8] Apollo raised technical challenges, but the greater challenge to NASA was to develop techniques for managing large projects and to forge an effective organization from its diverse group of independent-minded research laboratories.[9]

NASA succeeded in building a strong research organization, but the process resulted in another important ongoing tension. The laboratories the agency had inherited and even those it later created tended to develop their own independent interests.[10] Such was the case with the two organizations that contributed most to Landsat. The Johnson Space Center in Houston, Texas, established as the Manned Spacecraft Center in 1961 and renamed in 1973, concentrated its efforts on the Apollo lunar exploration program and on other plans for putting people into space. The Goddard Space Flight Center in Greenbelt, Maryland, created in 1958 as the Beltsville Space Center and renamed the following year, housed the agency's space science team. The use of spacecraft to study various scientific phenomena, particularly those outside the earth's inner atmosphere, brought scientists from many disciplines to Goddard. This center came to embody the scientific side of an ongoing tension between scientists and engineers at NASA.[11] Both centers had their own priorities, which did not necessarily agree with those of NASA headquarters.

Another question arose over the role of the space agency in routine operations in space. The space act specified that NASA was to be a research and development agency, and NASA leaders saw the agency's role as research and development of space exploration. The space agency necessarily conducted routine operations of some support technology, such as launch vehicles, but the agency structure emphasized research and development. NASA was primarily organized not around functions like development, launch, and mission control, but rather into teams for spacecraft projects ranging from Apollo to small scientific space probes. Because of this focus on research and development projects, the space agency had an interest in developing new technology for its own sake, rather than responding to user needs and producing the simplest system that would do any particular job.

Applications Satellites

NASA's spacecraft projects divided into three main program categories: people in space, space science, and applications. Putting people into space demanded a mammoth share of the agency's budget throughout the 1960s. Kennedy's mandate to reach the moon seemingly put the country in a space race with the Soviet Union that ensured the people-in-space program of congressional support and media attention until the feat was accomplished.[12] The scientific program, its beginnings inherited from the International Geophysical Year activities of the Naval Research Laboratory and others, served the interests of the scientists who supported NASA politically and helped to prove the peaceful intent of the U.S. space program.[13] The applications program—the use of space for practical purposes—grew out of scientific research and early military interests in reconnaissance, weather monitoring, and communications. The applications projects that NASA undertook during its first decade received less attention and support than people-in-space or scientific endeavors, largely because applications lacked a strong, vocal political constituency.

Applications satellites raised new issues for NASA because they involved research and development for users outside the space agency. When such spacecraft projects completed the process of research, development, design, construction, and testing and reached operational use—that is, when the technology stopped evolving and was used in a routine way for some practical purpose—NASA normally gave the management of the program to the agency that used the satellites. Thus, NASA developed the first weather satellites and continued to experiment with new technology, but the Weather Bureau funded and managed operational weather satellites, purchasing the necessary launch services from NASA. This relationship led to conflicts with users in the cases of all three applications satellite programs—weather, communications and earth resources.

NASA managers' early experience with meteorological and communications satellites demonstrated a variety of potential approaches that could be taken to the Landsat project. Political scientist W. Henry Lambright, among others, has pointed out that conflict often arises between developers and users when the agency developing a new technology is not responsible to the party that will actually use it.[14] That general description worked itself out in practice in many different ways, depending on the relative power of the groups involved, the divergency of their interests, and the number of different uses to which the satellites could be put. A number of specific issues that will be important to analyzing Landsat can been seen in the processes by which NASA turned weather and communications satellite projects over to their respective operational users.

Weather Satellites

The first of NASA's applications programs to reach the operational stage was the weather satellite program. In the case of weather satellites, tension between NASA and the prospective user, the Weather Bureau, arose only when the program was nearly ready to make the transition from an experimental to an operational system. The conflicts were clearly caused by lack of coordination between NASA and the user agency.[15]

Weather satellites carried television cameras that took images of cloud cover and then transmitted the images to earth. Two classes of weather satellites were developed: low-altitude satellites, which rapidly orbited the earth, taking images of various areas; and geosynchronous satellites, which orbited at such an altitude that they always remained over the same point on the earth's surface and therefore provided continuous monitoring of the weather on one side of the globe. The first weather satellites proved immediately useful for tracking hurricanes and other large-scale features difficult to observe as a whole from the ground. They also brought significant improvement of routine weather forecasting.[16] Such benefits became widely available in the early 1960s when the communications system included on NASA's Tiros weather satellites was improved so that the satellite could continuously broadcast the data it collected. These images could be received by any party with an inexpensive antenna and printer.

The first weather satellite, project Tiros (Television and Infra-Red Observation Satellite), originated at the Army Ballistic Missile Agency as a spin-off from reconnaissance satellite research. The project was transferred to NASA in 1959. During the development phase the Weather Bureau (a branch of the Department of Commerce) had little influence over NASA's program. The space agency did form an interagency advisory committee with the Weather Bureau, but this committee contributed little to the development of policy. The Bureau became heavily involved only in 1960, with the launch of the first experimental weather satellite, *Tiros 1*. NASA asked the Weather Bureau to analyze the data returned by the satellite. Meteorologists found the data very useful, and the Bureau became enthusiastic about satellites. Almost immediately, the Weather Bureau started using the data from Tiros to fulfull the Bureau's basic mission. The Bureau used satellite data to make cloud cover maps that it distributed to meteorologists to aid in developing routine forecasts (see figure 1). By the flight of the fourth Tiros in 1962, the Weather Bureau used data from the satellite operationally, providing daily transmissions of cloud cover maps to meteorologists.

NASA and the Weather Bureau disagreed about the next step after the successful Tiros experiment. NASA planned to follow the Tiros project with a more sophisticated series of proto-operational satellites called Nimbus. The Weather Bureau, however, found the Tiros data satisfactory and was suspicious of plans for Nimbus because it promised to be much more expensive than Tiros and might not be ready when the last Tiros satellite reached the end of its useful life. The Weather Bureau did not want to commit itself to

Figure 1
Meteorologist at the Environmental Science Services Administration (a precursor of the National Oceanic and Atmospheric Administration) prepares weather maps from photographs produced by a Nimbus satellite. (NASA, 1970)

a costly program of building and operating large satellites, which, once operational, would be paid for entirely from the Bureau's small budget. NASA did not respond adequately to the Bureau's concerns, and in September 1963 Weather Bureau officials took action, notifying NASA that the Bureau was withdrawing from the Nimbus program and the existing interagency agreement. Instead, the Bureau proposed the construction of a series of interim operational satellites based on Tiros and a new agreement that made NASA and the Weather Bureau equal partners. Faced with losing the whole program, NASA negotiated a new agreement with the Weather Bureau for a Tiros Operational System (TOS) but retained Nimbus as a NASA research and development project.

The Weather Bureau, a weak agency without much support from the Department of Commerce, could afford to threaten to stop cooperating with NASA only because the Bureau had found a powerful ally. The Department of Defense offered to work with the Weather Bureau and provide the necessary expertise with space technology if NASA refused to meet the Bureau's terms. Pentagon managers were concerned about the possibility of a gap between the Tiros and Nimbus projects, which would leave the military without the storm warning information on which it already depended. The Department of Defense was also motivated by jealousy of NASA for taking over projects from the military space program.

The weather satellite case provides a clear example of the tendency for a divergence of interests between research and development groups and users of new technology. NASA—the research and development agency—wanted to develop a second generation of satellites employing more sophisticated technology, whereas the Weather Bureau—the user—demanded the simpler, less expensive system already in hand and not yet fully exploited. Research and development groups tend to pursue their own interests even when such groups are part of an operating agency, and such behavior is to be expected in the case of an independent research and development agency. Had the research function been located in the operating agency the advance of new technology would probably have been slower, but the old technology might have been used more productively. Research and development groups tend to be concerned with hardware rather than with applications and to fail to see that more sophisticated technology is not necessarily better. On the other hand, independent research and development may result in improved technology more beneficial than the user had dreamed was possible.

Interagency politics had to provide the means to balance these conflicting interests. The Weather Bureau wanted one sort of satellite and NASA another, but instead of compromising or pleasing the customer, NASA initially preferred to ignore the bureau. This situation had to change if the Bureau was to manage the operational system. Because the Weather Bureau lacked bureaucratic clout, officials had difficulty getting NASA to agree to a more balanced arrangement, but eventually the Bureau found a way to get the necessary influence. The same questions of balance also arose in other NASA programs, particularly Landsat.

Communications Satellites

NASA's communications satellite program was another case in which problems occurred in the transition from an experimental to an operational system. Here, however, the governmental conflict centered on who would be the operational user. The communications industry saw the possibility of large profits in the development of such satellites, whereas Congress had to deal with tricky political issues of public versus private control of technologies developed by the government. The effect of the controversy on the technology was also different, showing the advantages of farsighted research and development.[17]

An overview of the technology shows a variety of technical choices that complicated the political choices. Communications satellites relayed radio waves carrying telephone, television, and data signals from one point on earth to another. NASA tested three varieties. Passive satellites, such as *Echo*, launched in 1960, simply provided a reflective surface (in that case a giant Mylar balloon) off which radio waves could be bounced (see figure 2). Both classes of active satellites received a signal from the ground, amplified it, and retransmitted it to its destination. Low-altitude active satellites, such as *Relay* and *Telstar*, moved rapidly relative to the surface of the earth. Thus, the earthbound transmitter and receiving antenna had to be pointed to follow the satellites, and a number of satellites were required so that one was always available above the horizon. Geosynchronous active satellites, such as *Syncom*, were placed in a geostationary orbit, where they circled at the same rate as earth turned, thus always remaining over the same point on the surface. This distant orbit (about 35,800 kilometers) required more powerful transmitters and more sensitive receivers on the satellite and at

Figure 2
A passive communications satellite similar to *Echo I* during ground inflation tests by engineers from NASA's Langley Research Center. (NASA, 1960)

the ground stations, but the fixed position meant that many fewer satellites were needed. In the end, almost all of the many operational communications satellites developed were of this type.

Many organizations had an interest in satellite communications. NASA started out with a limited role in this area of satellite research—first developing only passive satellites, then only low-altitude satellites—because of a November 1958 division of responsibilities with the Department of Defense that lasted until mid-1961. Unlike other applications programs, however, this type of satellite had an obvious potential for profit to private industry, which therefore played a role in research and development from the start. American Telephone and Telegraph (AT&T) and other smaller companies spent their own funds on communications satellite research in hopes of winning lucrative contracts later or, in the case of AT&T, in hopes of gaining a monopoly on a privately owned satellite communication system. AT&T developed its own experimental satellite, *Telstar*, and in 1961 announced that NASA had agreed to launch it. This plan would have put AT&T in a strong position to develop a commercial communications satellite system.

Yet, because of concerns by several government organizations about monopoly, diplomacy, and giving away the fruits of government research, private industry did not get the free hand it wanted in the communications satellite field. While agreeing to launch the AT&T satellite, NASA leaders insisted that a government-funded and -controlled experimental communications satellite also be built. On the basis of competitive bidding, NASA awarded a contract to RCA to build a low-altitude active satellite called *Relay*. NASA launched AT&T's satellite in July 1962 after the agency had awarded the contract for *Relay* but before its first launch.

NASA assumed that the government's experimental communications satellite programs would be followed by private-industry ownership of an operational communications satellite system, but this plan raised many questions. While NASA organized research, Congress fought over details of the institutional arrangements for an operational system. A number of different influences came into play. The Department of State feared that if a private company controlled the U.S. share of an international communications system, then that company would necessarily play a quasi-governmental role in negotiations with the government communications agencies of other countries. The liberals in Congress did not want to see the fruits of government research given away for private profit, whereas the conservatives wanted the government out of a function that private industry could handle. The communications industry (both domestic and international) and aerospace firms, of course, wanted as much of the control and profits as possible. The political process resulted in a compromise: The Communications Satellite Corporation (ComSat) was organized in 1962 as a private company with some board members appointed by the president, carefully defined federal regulation, and broad ownership by communications and aerospace firms and the general public. Even when the new corporation was established, questions about its actual function remained, and about two years of organization and system definition preceded the formal stock offering.

These political disputes slowed the development of the technology and altered its character. During the political controversy, NASA proceeded with research on a geosynchronous communications satellite, a more advanced concept too expensive and with too many technological risks for private companies to attempt on their own at that time. The 1963 flight tests of this satellite, *Syncom* (developed at Hughes Aircraft Company under contract to NASA), proved very successful (see figure 3). This changed many assumptions about communications satellites. For the first operational

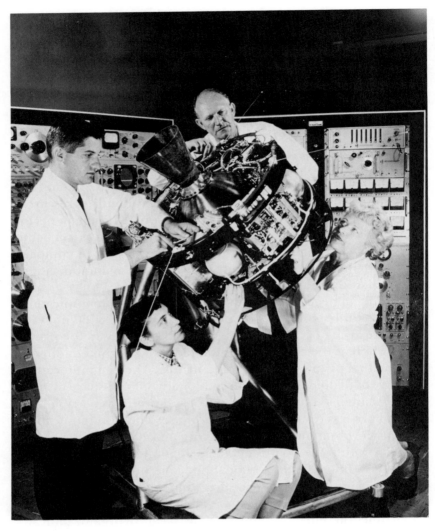

Figure 3
Checkout of the *Syncom 6* satellite by Hughes Aircraft Company engineers and laboratory technicians. (NASA, 1962)

communications satellite system, ComSat chose to develop not the system of low-altitude satellites that AT&T and the other communications companies had planned, but rather a system of geosynchronous satellites, initially called Early Bird. *Syncom* itself NASA turned over to the Department of Defense to avoid suspicions that the space agency might get into the operational communications satellite business.[18] The result of NASA research was a much less expensive commercial operational system because a geosynchronous system required fewer satellites.

This example shows the potential advantage of research and development that is not too tightly controlled by the users. In the case of communications satellites, the research and development agency did know better than the users. Unlike the Weather Bureau, the users of communications satellites wanted the more advanced technology that NASA had developed despite their initial lack of interest. The communications satellite case also shows that dealing with private industry as the operational users of an applications satellite system brings its own set of issues, on which there can be much political disagreement.

Earth Resources Satellites

Landsat proved to be a more complicated case than either the weather or communications satellite programs, but it involved many of the same issues. For earth resources satellites, NASA had to deal with a wide variety of users, leaving the overall goals of the program uncertain. Without a clear idea of who would use the satellite data and for what purposes, Landsat managers faced controversies with no easy resolution.

The technology for earth resources satellites was reasonably straightforward. Earth resources satellites provided wide-scale repeating series of pictures of the surface of the earth for the survey and monitoring of resources (see frontispiece).[19] The Landsat satellites, launched in 1972, 1975, 1977, 1982, and 1984, carried two sensors (discussed in chapter 6) that collected different kinds of images of the earth. The satellite radioed the data to ground stations, where it was printed on photographic film or analyzed by computers. Although the resolution was coarse, so that only objects larger than about 60 to 100 meters could be distinguished on images from the first two satellites, the satellites produced huge amounts of data. Processing, storing, and extracting information from the flood of data proved to be the most difficult technological challenge of the project (discussed in chapter 9). The data were used successfully, at least on an experimental scale, to detect large geological features associated with oil and minerals, to measure areas planted in different crops, to help predict harvests, to monitor water distribution and snow cover to predict flooding, and to make maps of land use. Users included many federal (chapters 11 and 12), state, and local (chapter 13) government agencies, and private firms (chapter 14).

A look back at the origins of the project shows that the federal agencies were the only users with any voice in the development of the first earth resources satellite. NASA established a program in 1964 to investigate how spacecraft might provide data on earth resources. The space agency immediately transferred funds to the Departments of the Interior and Agriculture so that scientists could consider what uses could be made of the kind of data such a satellite might provide. Interior developed so much enthusiasm for the idea that when NASA moved slowly in making plans for an experimental satellite, the department pushed the project along by announcing its own satellite program in September 1966. An independent satellite project was vetoed by the White House, since the development of experimental satellites was not in the domain of the Department of the Interior. This threat of intervention, however, caused NASA to accelerate its project.

To complicate matters further, there were many different requirements on Landsat. The Department of Agriculture proposed a different sensor from the one requested by Interior. Each department pushed for a small, simple satellite most useful to the applications that department wanted to develop. NASA compromised by flying both sensors and choosing their specifications to be useful for the widest possible range of applications. Later on some users complained that the specifications made all the data Landsat provided difficult to use because the data were not optimal for any one application. Compromise was also made in choice of orbit, and NASA settled for a satellite carrying the two sensors instead of a more ambitious project it had originally planned (see chapter 6).

Well into the program, NASA realized that more attention had to be paid to encouraging the wider practical use of Landsat. Some of the greatest potential benefits lay in improved resource management at the state and local levels. NASA had developed the satellite without consulting these potential users, and it proved difficult to persuade them to use the new information. The space agency established a program in 1974 to encourage the use of Landsat data, which the space agency referred to as technology transfer (although what was being transferred was not the satellite technology but rather the use of the resulting data). This program at first merely provided publicity for Landsat but gradually developed, with potential users, cooperative projects that were effective in convincing states to use Landsat data. Even with sophisticated technology transfer efforts, the use of Landsat data was adopted slowly because many resource managers distrusted sophisticated technology, which satellites and the space agency seemed to symbolize, and because they did not want to invest in using Landsat data until there was a fully operational satellite program.

The slow development of use of Landsat data after the launch of the first satellite in 1972 led the Office of Management and Budget to oppose the evolution of Landsat from an experimental project into an operational program. The commitment to an operational satellite system, to be managed by the National Oceanic and Atmospheric Administration (NOAA), was made only in late 1979. By that point competition for Landsat was appearing on the horizon, particularly a commercial French earth resources satellite project called SPOT, which launched its first satellite in 1986. The question of whether earth resources satellites should be owned and managed by the government or by private industry proved to be particularly hard to resolve. The Carter administration decided to give the whole project to private industry, but that proved to be hard to achieve because private industry did not see Landsat as profitable. In 1985 Landsat was finally turned over to a private corporation, but it is not yet clear whether that venture will be a success.

Landsat thus raised a more difficult set of questions about government technology for the public good than the other applications satellite programs. The tension between research and development was similar, but for Landsat it was much less clear who the ultimate users were and who should manage the ultimate operational system.

The comparison of NASA's applications programs shows that Landsat shared themes with other projects. Two common issues are particularly apparent. First, in each case a decision had to be made about who would manage the operational program. Second, there was a tendency for NASA to give the ultimate user little voice in the definition of the research and development program. This had both advantages and disadvantages, but it tended to make the process of turning the system over to the operational user more difficult. Landsat was an extreme example of these problems, so that project cannot be taken as typical. However, Landsat does raise issues with wider applicability.

3
Sources of Interest in Earth Resources Satellites

Interest in a civilian earth resources satellite arose late, after weather, communications, and classified reconnaissance satellites had already proved their usefulness. Interest in a civilian satellite came partly from a community of experts involved in research on Department of Defense reconnaissance satellites and partly from public enthusiasm for pictures of the earth taken by weather satellites and astronauts. This origin raises two issues. First, it is significant that Landsat was not based, except in the most general way, on reconnaissance satellite technology; one might have expected an earth resources satellite program to have used obsolete military technology. This did not happen because of tensions between NASA and the Department of Defense, tensions that eventually shaped the definition of Landsat. Second, it is striking that the development of a community of experts (and a new discipline) played such an essential role in the origins of a project otherwise dominated by bureaucracy. The germ of the idea for Landsat came from the ambitions of a scientific community, not from the planning process in the bureaucracy.

Aerial Photography

Landsat grew out of a long tradition of gathering data about the surface of the earth from aircraft and spacecraft. The development of aerial reconnaissance provided many of the fundamental techniques for both civilian and military satellites and also shaped expectations about how synoptic views of the earth would be used.

Aerial reconnaissance and photogrammetry—mapmaking from aerial photographs—have a long history. Aerial photography originated in the 1890s, using balloons as platforms.[1] In World War I, aircraft were used extensively for reconnaissance; in fact military strategists at first saw that as the primary value of aircraft.[2] After the war, the Army Air Service, searching for a peacetime mission, performed aerial mapping. Surplus military reconnaissance equipment fed a small civilian aerial photography industry, which provided photographs used primarily for mapping and geological exploration.[3]

Aerial photography took off in the 1930s. Planners of agricultural recovery efforts during the depression found that regional and national planning activities required current, high-quality maps of large areas, most conveniently made from aerial photographs. Agricultural Adjustment Administration engineers and technicians made and updated maps of agricultural land use so that the federal government could regulate production and locate abandoned farmland in need of erosion control.[4] Similarly, the Tennessee Valley Authority made extensive use of aerial mapping to help plan flood control work over a large region. This resulted in rapid growth of professions related to aerial map-making, leading in 1934 to the founding of the American Society of Photogrammetry and the journal *Photogrammetric Engineering*.[5]

The technological base in photographic aerial mapping developed in the 1930s proved useful in World War II. Photographic reconnaissance played an unexpectedly large role in the Second World War as a necessary adjunct to strategic bombing. The most significant contribution was made by a British group of photo-interpreters, led by Constance Babington-Smith, who identified targets, particularly military installations, for bombing and evaluated the results (occasionally to the dismay of the pilots). Their work showed the value of the trained eye in extracting information that a casual observer might not notice when looking at the same photograph.[6] The war brought developments in technology as well as in interpretation, most notably improved cameras and chemical flash bombs for nighttime reconnaissance.[7] Some research also started on infrared film capable of detecting camouflage.[8]

After the war ended, research and development of aerial reconnaissance hardware and techniques continued. Camouflage-detection work led to the development of color infrared film, which proved to have a wide range of applications. Infrared light hitting the film was translated into red in the resulting color picture, creating a false-color picture in which healthy vegetation, which reflected near-infrared light, appeared red instead of green. Vegetation cut for camouflage or damaged by insects, disease, or drought lost this reflectivity in the infrared before any change in color was visible in a normal photograph. Other research resulted in cameras with longer focal lengths that made it possible to take photographs showing significant details from higher altitudes, leading to overflights of the USSR by high-altitude balloons and, starting in 1956, by U-2 high-altitude reconnaissance aircraft.

More important to the Landsat story, research was also begun on nonphotographic sensors. Military strategists of the 1950s emphasized the importance of bombing, which led to research on using radar images for reconnaissance. These radar images had the advantage that during a mission they could be directly compared with the images on the radar that the crew of a bomber used for navigation and target selection. Military planners were also interested in sensors that could detect the heat emitted by people and machines; this required nonphotographic sensors to detect longer-wavelength thermal infrared radiation, which did not register even on infrared film.[9]

By the 1960s, then, aerial photography was both a well-established technique for routine applications and an area in which new technology promised continued progress. This had two effects on Landsat. First, it meant that a community of experts and research teams in industry was already established, making it easier for interest in earth resources satellites to develop. It also meant, however, that most of the available experts used their experience with aerial photography as the basis for their expectations of earth resources satellites. At minimum this limited their expectations of how the satellite data would be used (see chapter 9). At maximum, some experts in the field attached high importance to the areas in which aerial photography had advantages over earth resources satellites (fine resolution and flexibility of coverage) and argued that such satellites were not worth building because aerial photography gave superior results (see chapter 7).

Reconnaissance Satellites

Reconnaissance satellites evolved logically from other reconnaissance systems used by the military, but Landsat did not evolve directly from reconnaissance satellites. The capabilities the space agency developed for Landsat paralleled ones the Department of Defense already had, but NASA had to develop new technology because the Pentagon refused to share even obsolete classified technology. The Department of Defense also made it more difficult for NASA to get funding for Landsat (see chapter 7). The classified reconnaissance satellite program made its greatest contribution to Landsat by creating a community of interested and knowledgeable scientists and engineers.

The Department of Defense became interested in satellite reconnaissance immediately after World War II. Experiments with V-2 rockets captured from the Germans at the close of World War II and launched by the military in the United States proved the potential of reconnaissance from space. The navy began studying the practicality of satellites in 1945, and the RAND Corporation, an air force think tank, performed a number of studies of space vehicles in the late 1940s and early 1950s. As early as May 1946 a RAND study of "Preliminary Design of an Experimental World-Circling Spaceship" suggested reconnaissance as a primary mission for satellites, and an April 1951 RAND report analyzed the "Utility of a Satellite Vehicle for Reconnaissance."[10] Satellites had clear advantages for reconnaissance over the aircraft and balloons already in development and use because an orbiting satellite would be much more difficult to shoot down.

Actual development of a reconnaissance satellite system began in 1955 when the air force requested proposals for a "Strategic Satellite System." A contract was awarded to Lockheed in 1956, and the pace of the work accelerated after the launch of the first *Sputnik* in October 1957.[11] Although President Eisenhower transferred many military programs to the civilian National Aeronautics and Space Administration when it was established in 1958, the Department of Defense kept a tight hold on reconnaissance satellites. Military emphasis on reconnaissance satellites grew both out of direct military need and out of the expansionism of the military services. Walter McDougall has shown that starting in the early 1950s the highest councils of the U.S. government saw satellite reconnaissance as a crucial need, whose importance was further proved by the shooting down of an American U-2 reconnaissance plane by the Soviet Union in May 1960.[12] The services—the air force in particular—also had not given up hope of a large military space program, but the only practical use of orbiting vehicles that they could justify as having a clear military value was reconnaissance.[13]

Early reconnaissance satellites used different technological approaches from that ultimately used in Landsat. The air force initially investigated taking pictures on photographic film and bringing the film back to earth but concluded that the technical difficulties were too great. The Central Intelligence Agency picked up the system rejected by the air force and developed the first series of military reconnaissance satellites, part of a mixed military-civilian project called Discoverer.[14] These satellites took photographs of the earth during a few days of orbits and then ejected the film in a special capsule. The capsule reentered the atmosphere, descending by parachute, and was recovered in midair by an airplane towing a device that snagged the lines of the parachute. The first successful midair recovery of a Discoverer capsule occurred in August 1960, and the system was routinely supplying useful photographs within a few months.[15] The Discoverer program led to further generations of film-return satellites, with eighty-six vehicles placed in orbit by 1976.[16]

Film-return technology was not used for civilian satellites (for both security and technical reasons), but the air force meanwhile sponsored the development of another reconnaissance satellite technology that had more potential for nonmilitary missions. This kind of satellite processed the photographic film on board the satellite, scanned it, and radioed the resulting signals to earth. This had the advantage that data could be collected before the end of the useful life of the satellite, resulting in more timely data and longer-lived satellites. Such on-board processing was used on the air force *SAMOS* (Satellite Military Observation System) satellite, which had its first successful launch in January 1961.[17] NASA used similar technology on the *Lunar Orbiter* spacecraft to photograph the moon. However, NASA was not allowed to use a modified *Lunar Orbiter* to photograph the earth because Defense officials feared that such high-quality pictures of the earth would arouse interest in the capabilities of classified military reconnaissance systems, even though the Lunar Orbiter system gave coarser resolution than classified systems.[18]

Although none of this early military reconnaissance technology was directly used in the Landsat project, it is harder to determine whether more recent military technology was transferred. In 1961, when the Soviet Union criticized the United States for using space for military purposes, the Department of Defense stopped releasing any information about its reconnaissance satellites. What evidence is available suggests that the Pentagon had only a minor interest in the two kinds of sensors launched on Landsat in 1972: scanners and television cameras.[19] Infrared scanning and multispectral imaging were introduced onto reconnaissance satellites in 1966, but these were systems designed only for coarse-resolution surveys of large areas.[20] Television cameras were probably in use even earlier to provide broad, coarse-resolution surveillance.[21] However, these systems probably had even coarser resolution than the television cameras and scanners developed for Landsat. Instead of developing such systems further, the Pentagon sponsored the development of an alternative technology, a system to produce fine-resolution real-time imagery by capturing the image focused by a lens on an array of detectors called a *charge-coupled device* (CCD) instead of on a piece of photographic film. This system probably first came into use for classified reconnaissance in a new generation of reconnaissance satellites that had its first launch in late 1976.[22]

In general, the Department of Defense had only limited interest in those types of sensor technology used in Landsat because they provided only coarse resolution. Photographic systems used on reconnaissance satellites probably gave a resolution finer than thirty centimeters, compared with one hundred meters for *Landsat 1* and kilometers for weather satellites.[23] Fine resolution was essential for reconnaissance satellites because resolution determined the approximate size of the smallest square object that could be distinguished in an image.[24] Although there was much argument about exactly how much resolution was needed for earth resources satellites, extremely fine resolution was not desirable because most of the patterns of interest were large and because fine-resolution data were very expensive to process.

The Department of Defense discouraged the transfer of the technology it had developed to a civilian earth resources satellite program because the Pentagon did not want to risk damaging the secrecy that protected a technological lead in reconnaissance satellites that it believed to be crucial to national security. Defense refused to declassify even obsolete technology and restricted NASA to resolutions coarser than about twenty meters.[25] The Pentagon's refusal to share technology obviously made the development of earth resources satellites more difficult, but it may in the end have been an advantage. According to some individuals familiar with both programs, the unavailability of Department of Defense technology did not limit Landsat because most of the classified technology was not particularly well suited for studying earth resources.[26]

Different goals required different sensor designs: Most military surveillance concentrated on separating objects such as tanks and ships from the background and identifying them as exactly as possible. On the other hand, geologists were interested in the broad patterns of what to the military analysts was only the background, and agricultural analysts wanted to see seasonal changes in that background. More serious for Landsat were Pentagon worries that a civilian satellite, even using different technology, would call attention to the classified program. Reconnaissance satellites had been accepted only by precedent, not by international law, so there was still a danger that countries would object to being photographed.

The Scientific Community

Although military technology was not transferred directly, the civilian earth resources satellite program benefited from the expertise created by Department of Defense projects. Postwar military interest in a variety of sensors resulted in the expansion of the field of photogrammetry into a new research area called *remote sensing*. Military service and military contracts brought together a community of scientists interested in observing the earth with sensors other than traditional photographic cameras. Remote sensing scientists defined their field as the study of any method of gathering information from a distance, but in practice they usually limited their research to monitoring—and, particularly, making images from—various kinds of radiation, from light to radio waves. Many of these scientists, particularly the geologists and geographers, became interested in civilian as well as military applications of the technology they were developing. The remote sensing scientists who worked on research for the Department of Defense were the main source of early interest in a civilian earth observation satellite to monitor natural resources, and they played an important role in the program thereafter. However, they often lacked experience in dealing with information users other than the defense community.

Applied scientists in the new field of remote sensing found many opportunities for research. In the early 1960s the Department of Defense contracted with university laboratories for basic research on a variety of new sensors. Many of these contracts led to important innovations in remote sensing, and one of them played a direct role in the development of Landsat: a contract to the Infrared Laboratory of the Institute of Science and Technology at the University of Michigan (later the Environmental Research Institute of Michigan) for the development of battlefield surveillance techniques. This research effort, called Project Michigan, initially sought to develop sensors to detect tanks and people, but it later expanded to include ways of gathering information about terrain and ground cover—research with clear civilian applications (see chapter 5).[27]

Many scientists working on such contracts for the Pentagon became interested in the potential of a civilian remote-sensing satellite, and their interest led to the initiation of a series of conferences. In February 1962 the Infrared Laboratory held the first Symposium on Remote Sensing of Environment in Ann Arbor, Michigan. Funding was provided by the Geography Branch of the Office of Naval Research, which shared with many of the academic scientists an interest in pushing remote-sensing technology into the civilian arena.[28] Dana Parker of the University of Michigan gave the first paper, introducing the subject with the statement, "Modern aircraft and earth satellites now make it possible to survey the earth from a vantage point heretofore unobtainable."[29] The papers gave the impression that at least some of the speakers had access to photographs from reconnaissance satellites; certainly, many of them were using data taken by classified sensors on aircraft. The scientists wanted to develop civilian uses for such technology; this first conference focused on how to obtain the declassification of information and sensors they wanted to use. The attendees also formed working groups divided by scientific discipline to exchange ideas on the potential uses of remote sensing.[30]

Extrapolating from their military experience and reconnaissance research, the participants in the conference had a clear vision of a civilian remote-sensing satellite system. They argued, in fact, that they had already done much of the research necessary to prove the concept. At the second Conference on Remote Sensing of Environment, held in October 1962, participants concentrated on presenting existing work in the field in unclassified form. The keynote speaker, L. H. Lattman of Pennsylvania State University, explained what he saw as the value of the conference:

Equally important, it will demonstrate to all concerned that despite security restrictions there has been a great deal of progress in the field of remote sensing of the environment, that a great deal of work has been done and is available despite security restrictions, which certainly is an indication of what might happen if the security restrictions were handled more judiciously. From under the total blanket we are about to bring up some gems as an indication of what really does lie under that blanket.[31]

Another scientist, from the University of Washington, called specifically for a "civilian reconnaissance multi-sensor satellite."[32] He recognized the key advantages of satellites for remote sensing: repeated coverage "to produce virtual time-lapse photography" and coverage of much larger areas than could conveniently be achieved from aircraft.[33]

The interest of these scientists, their general knowledge, and their informal communications network were the greatest influence of the military reconnaissance program on what was to become Landsat. However, the cold war atmosphere of the early 1960s restricted the development of a new field of science using instruments already developed for military purposes.[34] The Pentagon refused to allow declassification of reconnaissance satellite technology, and Department of Defense concerns limited even the development of separate civilian programs. The remote-sensing scientists initially met little encouragement at NASA because NASA people avoided trespassing on the Pentagon's turf. Of the 186 participants in the first two remote-sensing conferences, only one was employed by NASA. The existence of a community of experts set the stage, but the play could not start until there was enthusiasm at the space agency.

NASA Programs

Enthusiasm at NASA finally developed as a result of other programs that hinted at the potential of looking at earth from space. Pictures taken by astronauts and weather satellites excited scientists and managers and provided those scientists already interested in the field with unclassified images with which to illustrate the many potential benefits of an earth resources satellite.[35]

The simplest kind of earth observation from space was performed by NASA's astronauts with their eyes alone. Looking at the earth from orbit during the last Mercury flight in 1963, astronaut L. Gordon Cooper, Jr., reported that he could see trucks on dirt roads, a train on its tracks, and buildings with smoke coming from them. At first the staff of mission control worried that Cooper might be hallucinating because their calculations showed that he could not possibly see an object that measured less than 150 feet on a side. However, the calculations turned out to be wrong. Scientists later found that Cooper's observations were made possible by three factors not taken into account in the first analysis: Cooper's vision was better than the 20/20 standard; the human eye can see linear features much narrower than the smallest square feature that is visible; and the loss of resolution due to the atmosphere was less than had been assumed.[36] The blurring effect of the atmosphere comes primarily from the air in the first few meters in front of the lens or the eye. Therefore, an instrument in space (outside the atmosphere) can see the surface of earth more clearly than an instrument on the surface can see objects in space.[37] This third factor not only helped explain Cooper's observation but also helped show the potential of satellite observations of the earth.

Public interest in the view seen by astronauts even before Cooper prompted NASA to include photography experiments on flights carrying astronauts. Hand-held cameras used on *Mercury 8* and *9* in October 1962 and May 1963 produced 50 to 60 photographs, showing much less detail than Cooper saw but still useful to scientists investigating earth resources surveying.[38] More systematic photography with hand-held cameras on *Gemini*, in 1965 and 1966, produced approximately 1100 photographs in normal color and color infrared useful to geologists (see figure 4). These photographs led to great public excitement, a few geological discoveries, and a wider scientific interest in remote sensing from space.[39] For example, the conclusion of a geological report on the Gemini photographs noted that "the experiment had totally unexpected benefits":

First among these was the great stimulus given to the NASA earth resources program and to remote sensing in general (from spacecraft and aircraft). It is not widely known that before the Mercury and Gemini photographs became available, there was virtually no appreciation of the potential value of orbital surveys for the management of earth resources.[40]

Gemini also gave NASA scientists experience with finer-resolution pictures of earth; in a classified experiment Gemini astronauts took photographs for the air force through a Questar telescope, which should have been capable of resolution of a few feet.[41]

The program to put people in space was not the only NASA program important to the origins of earth resources satellites. While pictures taken by astronauts were generating enthusiasm for earth resources photography, weather satellites provided NASA with a technological and system model. As early as 1963, scientists tried to use weather satellite images for earth resources investigations, but the very coarse resolution made the data of little use.[42]

Figure 4
Gemini photograph of the Imperial Valley of California and the Joshua Tree National Monument area. (U. S. Department of the Interior, Geological Survey)

More important was the model: Weather satellites utilized television cameras that provided repetitive pictures of the scene below and radioed the images promptly to earth, where they were widely distributed. This was one model for an earth resources satellite; in fact, some early proposals suggested putting a sensor suitable for monitoring earth resources in an existing series of weather satellites.[43] Although such specific links between meteorological and earth resources satellites were not developed, many of the issues that arose for weather satellites were important to Landsat (see chapter 2).

By 1963 or 1964 all the elements existed that were necessary for assembling an earth resources satellite program. Mercury observations and photographs had shown NASA the potential of looking down at earth and had created interest in studying earth from space. A community of remote-sensing scientists familiar with much of the necessary theory and with some idea of the possible civilian uses of satellite data provided the necessary technical base. And NASA's successful weather satellite program provided a model for a similar effort in earth resources. Unfortunately, these origins contained not only great potential but also the seeds of future problems. Serious tension would arise between NASA and the military over protecting classified technology. More subtly, problems were to arise because the earth resources satellite program grew out of recognition of technological potential, not user demand. The resource managers who would eventually use the new information had no role in the origin of the program.

II
Development

4
NASA Establishes a Program

From 1964 to 1966 NASA managers and engineers discussed alternative approaches to initiating a program to develop technology to monitor earth resources from space. They proposed a variety of long-range plans for earth resources observation from space, but the agency did not officially make a commitment to a particular satellite project in this period. Instead, top NASA managers added earth resources observation experiments to plans for spacecraft carrying people, particularly the space laboratory, *Skylab*. What discussion there was of an eventual earth resources satellite without a human crew centered on testing a wide variety of sensors on a single experimental payload rather than building the simplest possible useful satellite.

In this early stage of the evolution of the project, conflicting interests within NASA could be seen particularly clearly. When NASA established an earth resources program in 1964, many possible technologies awaited investigation. Choices made among these options in early planning for earth resources observation from space reflected the interests of various groups within the space agency. Overall, NASA leaders were interested in developing space technology for its own sake; they therefore emphasized lengthy experimentation and technological sophistication. They also used the potential of observing earth resources from space to support an active program to put people into space. The first head of the earth resources program, meanwhile, emphasized fundamental research, not planning for a useful satellite, and a research team at the Johnson Space Center emphasized development of sensors.

Space Science For Apollo

NASA's earth resources program originated in the organization developed for the Apollo project at NASA Headquarters. The Manned Spaceflight program included an Office of Manned Space Science until late 1963, when that office was transferred to the Office of Space Science and Applications. Even in its new location, the Office of Manned Space Science continued to receive funding from the Manned Spaceflight budget and preserved a different approach to space science from that typical of the Office of Space Science and Applications. The style of most of NASA's space science program had been set by an early emphasis on studying particles and fields in the space near the earth; exploring the planets came to dominate the agency's space science program only later.[1] The Office of Space Science and Applications in the mid-1960s was led by astronomers and physicists interested in particles and fields rather than by planetary geologists. The Office of Manned Space Science, on the other hand, hired and supported a large group of geologists to plan for and analyze the results of lunar exploration. These geologists provided a new core of interest in the earth at NASA.

Geologist Peter C. Badgley came to NASA in June 1963 from the Colorado School of Mines. As program chief for advanced missions in the Office of Manned Space Science, his task was to improve the scientific program for Apollo.[2] As part of this planning, he investigated the possibility of testing systems for observing the moon (like the one ultimately used on *Lunar Orbiter* in 1966 and 1967) by looking at earth from orbit. Although that plan foundered, Badgley became increasingly interested in techniques for observing earth from space.[3]

Badgley made earth observations one of the scientific goals in his long-range planning for spacecraft to carry people. He obtained the approval of his superiors, apparently in December 1963, to sponsor investigations focused specifically on the study of earth resources from space. During fiscal year 1964 Badgley funded studies of the potential of remote sensing of natural resources at the Department of the Interior, the Army Corps of Engineers, and a few university laboratories.[4] The results were promising. These early studies provided evidence of the value of using spacecraft to observe earth resources, defined some of the key scientific problems, and involved organizations that were to play a large role in the Landsat project.

As early as the fall of 1964 Badgley had a specific program of earth resources studies in mind as part of his planning for Manned Space Science. The list "Earth-Oriented and Lunar-Oriented Experiments for Manned Orbital Missions," compiled in the fall of 1964, included—as three of the five user objectives—oceanography, worldwide forecasting of agricultural conditions, and studies of natural resources in developing countries. The author, presumably Badgley, considered the possibility of using a small automated satellite but concluded that sensors with sufficient resolution would be too heavy for a satellite similar to the Nimbus weather satellite. He argued that a large satellite should include a human crew, because "man's ability to select the most interesting targets for detailed study is essential."[5]

Research Contracts

Badgley contracted for research in a number of fields based on his view of the objectives of a NASA earth resources program. Initially, most of this was exploratory research, not even developing sensors but investigating the parameters that would determine what kinds of sensors should be developed. Later the focus shifted to scientific research on a wide variety of potential sensors for use in many different disciplines.

In agriculture Badgley let contracts to the University of Michigan and Purdue University for studies of the possibility of identifying crops by remote sensing in order to forecast harvest size (see chapter 12 for the later development of this research area). Early work by the Department of Agriculture, using aerial photography, had shown scientists that differentiating crops would not be easy. At many stages of growth, small grains such as wheat, barley, and oats are almost identical in color. However, if the color of the plant is broken into a spectrum, showing how much of each color of light the plant reflects, a characteristic pattern called a *spectral signature* emerges. Such research started from laboratory studies of the spectrum reflected and emitted by different leaves.[6] Since an airborne or spaceborne sensor could not usefully measure so detailed a spectrum, scientists then had to choose a few wider regions of the spectrum—*spectral bands*—that sensors could measure to differentiate between various crops in the field. The University of Michigan had developed sensors that could be used to test this concept (discussed further in chapter 6). NASA supported research involving both detailed study of plant leaves in the laboratory and tests of these sensors from aircraft over known fields.[7]

In geology Badgley funded a variety of studies aimed at correlating the appearance of the surface of the earth with phenomena of interest beneath the surface, such as oil deposits. Aerial photographs clearly showed some large-scale geological features, such as faults. Research was needed to correlate these features with resources such as oil and minerals.[8] As with crops, scientists hoped to identify different types of rock by their spectral signatures, but this proved to be even less successful in geology than in agriculture (the agricultural case is discussed further in chapter 12, the geological in chapter 13). Other types of sensors also held promise for geology. The Universities of Kansas and Michigan received contracts to explore the use of radar to investigate geological features from aircraft,[9] and Ohio State used radar systems mounted on trucks for more detailed studies to try to understand how different materials reflected radar waves.[10]

After determining that remote sensing had potential for a wide range of disciplines, the next step in preparing for a satellite project was to research appropriate sensors. In particular, NASA encouraged the development of nonphotographic sensors. Most of the scientists interested in earth resources observations from space were already involved in research using data from photographic cameras with infrared film and did not necessarily see a need for new sensors. NASA scientists and engineers wanted to encourage the use of other sensors, however, both because the Department of Defense was not likely to let NASA use photographic cameras in Landsat and because they wanted to pursue more advanced technology. The technologies that NASA engineers had the most interest in, scanners and vidicons (see chapter 5), had the advantage that they could be combined with filters to generate data from narrow spectral bands in many parts of the spectrum. An infrared photograph, on the other hand, provided information about one broad region of the spectrum, the size of the region depending on the sensitivity of the photographic emulsion. The approach taken by NASA scientists and engineers had real advantages, but its technological sophistication made it harder to get outsiders involved.

The data provided by these sensors required extensive research to learn to use its full potential. For example, NASA contracted with the Environmental Research Institute of Michigan for further studies of a scanner the institute had developed.[11] These NASA-sponsored studies provided information useful for the development of sensor technology, as discussed in chapter 5. In addition to developing sensors, researchers found they faced a significant challenge in learning to interpret the data. The Michigan studies provided data from a variety of terrains and land uses so that the researchers could determine what phenomena showed up most clearly in the scanner images and learn what various conditions would look like in the images.

Before Landsat data could be used on an operational basis, researchers had to determine basic "ground truth." In other words, data from a specific image had to be correlated with the actual features on the ground. For example, to learn to differentiate wheat fields from corn fields in an image provided by sensors carried on a high-flying aircraft or spacecraft, scientists first identified wheat and corn fields in an image of a known area where actual ground surveys had been conducted. Researchers had to determine not only the spectrum of light a particular kind of vegetation reflected but also how this spectrum varied with the growth of the plant, the variety being grown, the character of the soil showing between the rows, and regional variations in moisture and soil. Even the orientation of the rows of a field crop relative to the scan lines of the sensor affected the appearance of a planted field in an image taken by a scanner like that on Landsat.[12] Once scientists worked out methods of differentiating various kinds of vegetation, they still had to develop methods to use the resulting information to predict harvests or to support land-use studies.

Badgley used experts both inside and outside NASA to conduct this program of scientific research. Starting in April 1965, the geological studies were supervised by Badgley's deputy chief, Alden P. Colvocoresses, a geologist on loan to NASA from the army. Colvocoresses played an important role in developing the cartography program for earth resources satellites. He also acted as a gadfly, first in his position at NASA and then, starting in 1968, at the U.S. Geological Survey.[13] To supplement the expertise of insiders, Badgley organized advisory teams of university scientists in such disciplines as geology and forestry. Many of these scientists had been introduced to the general field of remote sensing of the earth by Department of Defense research contracts (see chapter 3). For example, the head of the forestry team, Robert N. Colwell of the

University of California at Berkeley, had been using infrared photography for forestry research since the 1950s in various classified projects.[14] Although these scientists wanted to put their research to practical civilian use, they sometimes had no more knowledge than NASA of the needs of the resource management agencies who would ultimately use the techniques they developed.

Aircraft Program

NASA's enthusiasm for developing sophisticated technology showed up immediately in the agency's own remote-sensing research program. In addition to contracting with universities for studies of remote sensing of earth resources, Badgley provided funding for the Johnson Space Center in Houston to establish a research program for testing sensors from aircraft. The head of this program was Leo Childs, an enthusiastic pilot who played a large role in the earth resources aircraft program until his retirement. This project tended to take on a life of its own as a research project to investigate sensors from a variety of aircraft platforms; it resulted in new technology for aerial survey (see for example chapter 13) as well as in information needed to develop a satellite remote-sensing program.

The issue of the relative value of aircraft and spacecraft for earth resources surveys later became controversial (see chapter 6), but aircraft studies were clearly required to prepare for observations from space. NASA engineers and scientists used aircraft to test earth resources sensors because flying a sensor on an aircraft was, in the early testing stages, much less expensive than flying it on a spacecraft. The sensor could be repaired or improved between flights, and aircraft could carry heavier and larger instruments than could satellites; aircraft could also provide the instruments with more electric power.[15] A range of platforms was necessary even within the aircraft program; Childs's group started by flying sensors at a lower altitude because that simplified analysis of data provided by a particular sensor by reducing the effects of atmospheric absorption and making it easier to achieve fine resolution.[16] Later, aircraft at intermediate and high altitudes were used to test sensors under more realistic conditions and to provide sample data before the launch of a satellite, because such aircraft provided the same sort of broad view as a satellite.

The aircraft research involved a variety of sensors and aircraft to carry them, with an emphasis on developing sophisticated sensors. Childs's group acquired its first aircraft, a Convair 240A, in late 1964. At first this plane was used to test Apollo lunar radar, but the emphasis of the program quickly changed to earth resources.[17] For study of the earth, the Convair carried mapping and multispectral cameras and, later, radar and infrared systems, many of them obtained from the military.[18] By mid-1965, plans were underway to obtain a P3V aircraft that could operate at intermediate altitudes (7000 to 11,000 meters). Childs's team also acquired a high-altitude aircraft, a WB-57, in 1969.[19] The Earth Resources Aircraft Program did some of the same kinds of research as scientists supported by NASA at universities, but it tended to emphasize photography in the visible and near-infrared regions of the spectrum as well as radar, whereas the University of Michigan group developed the multi-spectral scanners ultimately used on Landsat.[20] But Childs's team was interested in pushing the state of the art in all directions; they even experimented with a twenty-four-spectral-band scanner that gave more data than anyone knew how to use.[21] The aircraft group provided data from the various sensors they tested to investigators at the Departments of Agriculture and the Interior and at various universities.

In addition to testing the advantages of different sensors, sensors carried on aircraft provided intermediate data necessary to understand satellite images. Aircraft provided data for user experimentation before the satellite was launched and findings to compare with the satellite data. Early pictures from space could be interpreted only with the help of a kind of a process called *multistage sampling*, which provided a kind of Rosetta stone. To learn to interpret satellite data, scientists compared satellite images with aircraft pictures of selected areas taken at the same time. By comparing the satellite data with aircraft pictures that they already knew how to interpret and with additional observations made on the ground, the experts refined the interpretation process. By a continuous chain of development, earth resources scientists worked from field studies to low-altitude aircraft photos to high-altitude aircraft photos to satellite images. Each step was built on the foundation of the previous one.

Development of applications became more central to the Earth Resources Aircraft Program as the launch of the first satellite approached. During fiscal year 1970 Johnson Space Center aircraft, carrying a variety of sensors, flew over seventy-two different test sites for investigators representing twenty-nine organizations.[22] NASA hoped that these exercises would build a constituency of Landsat supporters and give a large number of potential users the experience needed to make good use of Landsat data when it became available. In 1971 and 1972 preparation for Landsat dominated the earth resources aircraft program. NASA specialists produced simulated Landsat data from sensors carried on high-altitude aircraft, including two U-2 reconnaissance aircraft.[23] One part of this effort, the Corn Blight Watch, is discussed in some detail in chapter 11. Such aircraft studies proved to be a good way of answering some of the fundamental scientific questions of how to interpret remote sensing data. After the launch of the first Landsat satellite in 1972, NASA's earth resources aircraft program continued to test advanced sensing equipment to develop new technology and learn how to interpret the data provided by new sensors.

Satellite Plans

While the aircraft program in Houston tested sensors and developed ways of interpreting the results, Badgley's group at NASA headquarters started to plan to orbit those sensors. Because Badgley was in the Office of Manned Space Science, he emphasized the use of spacecraft to carry people. The planners of the program for human spaceflight expected a natural progression from hand-held cameras on Mercury and Gemini to more sophisticated cameras on Apollo and Apollo Applications spacecraft to a permanent space station with a significant earth resources mission.[24]

Badgley gave earth resources experiments a major role in his early planning for the post-Apollo human spaceflight program. During the initial enthusiasm for Apollo, NASA had planned to extend the project into a program of further studies of the moon and earth. These ambitious plans were derailed by a lack of political support for an extensive ongoing program once the race to the moon had been won. In response to the new climate, the Apollo Applications program was first scaled down to simply use up excess Apollo vehicles and finally reduced to a single laboratory spacecraft called *Skylab*, launched in 1973. Extensive earth resources experiments were included in early planning for Apollo Applications, but the mission (AAP 1A) designated to carry them was cut in late 1967.[25] In late 1969 congressional support for earth resources projects persuaded NASA to put a major package of earth resources experiments back onto *Skylab*. The package that was finally flown included fine-resolution cameras, tests of microwave sensors, and a ten-band multispectral scanner.[26] The data produced proved useful for scientific research but of little value for developing practical applications because it covered only a limited area and a limited time period.

The old controversy between NASA and the scientific community about whether it was better to use spacecraft with a human crew or automated space vehicles thus arose for earth resources as well. The scientific community preferred automated spacecraft because they were less expensive and more flexible, while NASA wanted to maintain a large role for people in space. NASA was accused of saving earth resources experiments to use as a justification for human spaceflight, particularly Apollo Applications.[27] Indeed, earth resources survey plans other than Apollo Applications experiments received little attention from NASA's leaders. However, this most likely reflected the slow rate at which NASA leaders came to recognize the political value of space applications and to believe that earth resources survey from space could be an important use of civilian satellites. A June 1966 survey article on "Space Science and Applications: Where We Stand Today" by NASA's Deputy Associate Administrator for Space Science and Applications did not even include remote sensing of earth resources.[28]

Badgley's emphasis on the use of spacecraft that carried people for earth resources studies probably arose more from technological momentum than any conscious intention to reserve earth resources as a justification for putting people in space.[29] Badgley worked for the Office of Manned Space Science, and earth resources research at NASA had begun with observations by astronauts. Badgley also justified the use of a human crew on the grounds that short flights that returned film and magnetic recording tape to scientists on the ground could efficiently test a variety of sensors for eventual use in a long-lived automated satellite. A summary of NASA's earth resources program presented by Badgley to the April 1966 Symposium on Remote Sensing of Environment projected that Apollo Applications flights would serve as the first tests of earth resources sensors in space. These tests would evaluate a wide range of sensors and provide the necessary information to decide whether the operational system should be an automated satellite, a space station with a human crew, or a spacecraft serviced by occasional visits by astronauts.[30]

Badgley did not neglect automated satellites entirely, although no plans were far enough advanced to be mentioned in an April 1966 presentation. Prior to the fall of 1966 NASA studied three classes of possible satellites for earth resources observation from space, generally called small, medium, and large orbiting earth resources observatories. These studies represented only an investigation of options; the Office of Space Science and Applications planning process normally involved the discussion of many possible spacecraft that were never built. Of the Earth Resources Observatories, only the small one evolved into an approved project, which eventually became Landsat.

As NASA began to explore these satellite possibilities in the spring of 1966, Badgley took an ambitious approach stressing the development of sophisticated experimental satellites. In May 1966 Badgley and his supervisor, Leonard Jaffe, prepared an outline of NASA's earth resources program that stressed the large satellite. Badgley and Jaffe called for orbital tests of a multispectral scanner and a photographic camera of approximately thirty meters resolution on a flight already planned.[31] Following this, in 1970 or 1971, they suggested that NASA launch an automated satellite carrying eighteen sensors: four camera systems, three radar systems, six infrared and microwave scanners and radiometers, and ultraviolet, laser, gravity gradient, and magnetometer systems. They argued that

flight test of this entire group of instruments simultaneously is highly desirable because the cross correlation of data from various portions of the electromagnetic spectrum acquired under similar lighting and weather conditions will yield far more information than data acquired at separate times by individual instruments.[32]

This complex experiment was typical of NASA's tendency to pursue technology for its own sake into extremely thorough research. Homer E. Newell, Jr., longtime director of space science at NASA, has pointed out that the Office of Space Science and Applications tended to plan more complex experiments than even the scientists wanted.[33]

On a more reasonable scale, a memo by Badgley in August 1966 stressed the medium orbiting earth resources observatory as the ultimate operational system. He suggested that two experimental medium-sized satellites, to be launched in 1972 and 1973, test sensors for the operational system. The instrument payload was to be similar in size to that of an Orbiting Astronomical Observatory (a scientific satellite with about 450 kilograms of instruments).[34] This class of satellites could carry sensors for all wavelengths but not the variety carried by the large satellite.

The same August 1966 memo suggested that a small orbiting earth resources observatory could meet requirements submitted by the U.S. Geological Survey. Such a satellite would be similar to Nimbus, carrying about 90 kilograms of instruments. The Geological Survey proposed that the satellite's payload include some combination of photographic, television, and infrared scanner sensors.[35] Because of the Survey's interest, Badgley placed this scheme on the official list of experiments under consideration by the Office of Space Science and Applications in early September 1966.[36] As will be seen in the next chapter, this was not enough to satisfy officials from the Geological Survey.

Except for this small satellite added at the request of the Geological Survey, Badgley's earth resources program emphasized large vehicles, particularly those carrying people, and extensive tests of a variety of complex sensors. This was a natural result of Badgley's position in advanced planning for space science in the human spaceflight program. This approach was also typical of NASA's style, but it was not conducive to the quick development of a basic useful system; therefore, it became a root of conflict between NASA and the organizations that hoped to use the satellite data.

5
User Agencies Seek a Role

The decision to build a particular type of earth resources satellite involved not only differing interest groups within NASA but also the interests of agencies that NASA identified as potential users. Initially, NASA involved future users of earth resources satellites in the Landsat program only to the extent of supporting research at a few federal agencies. Even this limited user involvement had a significant impact on the project; groups within those agencies attempted, although with limited success, to play a major role in defining the new program. Although the space agency depended on the user agencies for some scientific research, it did not always agree with them about what their role in the overall program should be.

NASA-funded experiments generated great enthusiasm among scientists at the Departments of Agriculture and the Interior about the potential of satellite data for agricultural and geological studies. In fact, the Department of the Interior became a strong lobbyist for an accelerated satellite program shaped to serve what it perceived as its needs. The resulting attempt to pressure NASA to accelerate its program reflected interagency power-playing, different ideas from those held by NASA and the Department of Defense about what an earth resources satellite should do and what technology it should use, and a sensitivity to the needs of at least a small part of the potential user community for Landsat data. Not surprisingly, NASA managers resisted this pressure, wanting to design the program with as little interference from outside as possible. However, the process of interagency politics enabled some compromises to be made among these conflicting interests. The question addressed here is not whether interagency politics reflected or subverted national policy but how interagency negotiations shaped definitions of the developing Landsat project, what technology would be used for Landsat, and how that technology would be used. Chapters 5 and 7 concentrate on specific political problems, whereas 6, 8, and 9 examine technological choices.

As discussed in chapter 4, one of Peter Badgley's first moves when he received authorization for earth resources research was to fund research at the agencies that could potentially use Landsat data to help fulfill their assigned missions. He transferred NASA research funds to the Departments of Agriculture and the Interior for agricultural and geological research and provided funds to the Naval Oceanographic Office to investigate using a satellite for oceanography. This last area of research was transferred from the Naval Oceanographic Office to the National Oceanic and Atmospheric Administration when that agency was created. It led eventually to *Seasat*, a satellite designed for oceanographic research, launched in 1978.

Contracting with other agencies for research was standard NASA procedure. The Army Corps of Engineers and the Geological Survey had already contributed special skills to the Apollo program and had been reimbursed by NASA, so it seemed logical to call on the Departments of Agriculture and the Interior for scientific expertise in the field of earth resources.[1] That NASA paid the departments for their services rather than bringing them in as partners in the enterprise shows that NASA considered itself solely responsible for space research and development. The departments found this arrangement useful because as operating agencies, they had difficulty funding their own advanced research. NASA's other goal was to enlist the political support of these potential beneficiaries for NASA's new applications program to help persuade the Bureau of the Budget and Congress of the value of the program.

Tension Between NASA and the Department of the Interior

Most of the funding provided to the Department of the Interior went to one of its branches, the U.S. Geological Survey, which used this funding to assign a group of scientists to investigate potential applications of remote sensing from space and to determine specifications for an earth resources satellite and additional experiments on Apollo Applications missions. In 1966 Badgley asked the Geological Survey to prepare a summary evaluation of the potential value of an earth resources satellite for geology, cartography, and hydrology. The Geological Survey scientists reported that remote sensing from space would be very valuable in these fields and should be developed quickly.²

By mid-1966 the group at the Department of the Interior became impatient with NASA's lack of progress toward defining a satellite system. The leaders of this discontented group were a geologist in the Geologic Division of the Geological Survey, William Fischer, and hydrologist Charles J. Robinove. NASA had been discussing earth resources experiments for the human spaceflight program and small, medium, and large orbiting earth resources observatories for at least a year without actually approving any project. The scientists at the Geological Survey believed that the experiments NASA was discussing were too diverse and sophisticated; the Geological Survey group wanted a simple useful satellite as soon as possible.

NASA leaders, in the meantime, had many reasons to proceed slowly with an earth resources satellite. Earth resources experiments for the Apollo Applications program also had to be developed, taking time and energy away from work toward an automated satellite. More importantly, NASA may have had a secret agreement with the Department of Defense or the National Security Council to go slowly on earth resources satellites.³ The details of this policy, if it exists, will not be available until information on reconnaissance satellites is declassified. There is, however, evidence of much discussion in late 1965 and early 1966 of what resolution could be used without classification problems.⁴ This suggests that the relationship between reconnaissance satellites and earth resources satellites was an issue at that time.

While NASA leaders worried about such policy problems, the people on the program management level at NASA, particularly Badgley and Leonard Jaffe, head of NASA's Applications program, were sensitive to the demand for the speedy development of an earth resources satellite. They continued to fund a range of earth resources research and tried to speed the program up by arguing to their superiors that NASA needed to reorient itself toward practical missions.⁵ By 1966 the program had a good foothold in NASA, although not a great deal of support from the agency's policy makers, and the NASA scientists were aware of growing interest outside the agency.

In an August 1966 memo to the record Badgley delineated these concerns. Arguing for a small orbiting earth resources observatory, Badgley pointed out that the user agencies wanted an operational satellite in 1970 instead of in 1974–75, as NASA officials intended. Badgley stressed that NASA should take a leading role in the program, including providing research funds to other agencies, because strong leadership was particularly important for a satellite that would collect data for many agencies. He pointed out that the benefits of studying earth resources from space could help justify the space program if earth resources experiments lived up to their promise and that applications "is an area which NASA is holding up to Congress as very promising, and one which needs strong support."[6] The same memo mentioned a fact that was to come as a shock a month later: The Geological Survey was considering asking Congress for money in fiscal year 1968 for construction of a small satellite, for a launch possibly as early as 1969. The Survey, according to Badgley, considered photographic and television sensors and the necessary communications system "to be already operational for space use" for needs in the fields of geology, hydrology, cartography, and geography.[7] Officials in the higher levels of NASA did not heed Badgley's warning.

An August 1966 meeting of representatives from NASA (including Jaffe), the Geological Survey (including Fischer and Robinove), the Agency for International Development, the Department of Agriculture, and the Naval Oceanographic Office clearly brought out the Survey's concerns. The Geological Survey and the Agency for International Development were enthusiastic about using data collected from space for resources studies in Latin America. This discussion had been partially motivated by a State Department paper that suggested studying the feasibility of using a spacecraft to collect earth resources data over South America.[8] Jaffe, with caution typical both of NASA's position and his own personality, warned "about overenthusiasm on [the] part of underdeveloped nations for new and exotic techniques, with concomitant exclusion of proven methods."[9] Fischer discussed the idea of a Department of the Interior Earth Resources Observation Satellite, based on an RCA proposal using a newly developed very-fine-resolution television camera (discussed in chapter 6).

This meeting only made the different interests more suspicious of each other's motivations. In a memo on the meeting Jaffe concluded that "USGS would like to budget for an 'operational' satellite as soon as possible to establish jurisdiction."[10] In other words, he believed that the Geological Survey was pushing earth resources satellites in order to ensure that it would control the eventual operational satellite system. While Geological Survey leaders were probably interested in grabbing a new project, Jaffe's description seemed to give little credence to the Survey's preference for a particular type of satellite. NASA's position, meanwhile, seemed to the user agencies to promise further delays. In the meeting Jaffe suggested that the user agencies should develop data requirements; NASA would then consider what sort of satellite would meet those requirements after "certain decisions" were made. Those decisions, presumably classified, were expected in September.[11] Jaffe's remarks implied a lengthy experimental phase under NASA's control, very different from the quick development of an operational satellite that the Department of the Interior wanted.

Interior's Announcement of Its Own Program

The Department of the Interior wanted to control an earth resources satellite program both because it believed that data from such a satellite would help the agency fulfill its mission and because, like most agencies, it sought to increase its own areas of responsibility and thus its size and power. By mid-1966 Fischer and Robinove at the Geological Survey had a good idea of what an earth resources satellite could do and who could use it. They were frustrated by NASA's rapidly changing ideas for a variety of earth resources satellites and lack of any approved project in progress. Fischer and Robinove wanted to force NASA to commit itself to building the small satellite they believed would be most useful.[12] Their first step was to go to William D. Pecora, director of the Geological Survey, with the idea that the Department of the Interior might develop its own operational earth resources satellite. In an interview, Robinove described why Pecora liked this idea:

Now Bill Pecora . . . recognized that this concept technically was one that if it worked . . . [would have] a very, very large payoff. It would be very worthwhile . . . and . . . [it] would be good for the country to have this information . . . . Also, being a very astute politician, he recognized that if we took the leadership, then the Geological Survey and the Department of the Interior stood to gain simply by being a good forward-looking agency.[13]

Fischer, Robinove, and Pecora presented the idea to the Secretary of the Interior, Stewart L. Udall, arguing that the satellite would be useful to other branches of the Department of the Interior in addition to the Geological Survey. Udall was quickly convinced of the merits of a Department of the Interior satellite program. According to Fischer, Udall may have liked the idea in part because he resented the amount of government money going to military and space programs rather than to the preservation of natural resources and undeveloped areas that he valued so highly.[14]

The enthusiasts at the Department of the Interior prepared a plan in secret, and most observers were shocked when Udall unilaterally announced on September 21, 1966, that the Department of the Interior planned a program of Earth Resources Observation Satellites (EROS). At a news conference, Pecora described EROS as an evolutionary program leading to a first launch in 1969 of a satellite that would carry "television cameras flown in an orbit that will cover the entire surface of the earth repeatedly."[15] Although the proposed satellite had many of the features eventually incorporated in Landsat, the announcement appears to have been largely a political move designed to accelerate NASA's program. In fact, as William Fischer remembered one reaction to the announcement, "Pecora had no more resources than the Prince of Liechtenstein with which to launch that satellite."[16]

Reactions to the EROS Announcement

NASA leaders were upset over Udall's announcement, nonetheless.[17] The threat they saw was typical of interagency conflict: Interior appeared to be attempting to take over a function that clearly belonged to NASA, the development of a new satellite system. NASA's only mission was research and development; if other agencies developed their own space systems, they would take away NASA's reason for being. Fischer remembered one NASA official telling an Interior official, "If you so much as used a nickel of NASA's money to write that press release I'll see you in jail."[18]

The negative reaction was not limited to NASA. Willis H. Shapley, who had moved from the Bureau of the Budget to NASA in September 1965, reported that Udall was nearly fired because the Department of the Interior announcement was in conflict with national policy so secret that Udall had not been informed of it.[19] Pecora said a few years later that he too had feared he would lose his job.[20] In retrospect, however, the reaction was not entirely hostile. Many people at NASA remembered the EROS announcement as a "good friendly bureaucratic maneuver to get the guys moving and doing something instead of just talking."[21]

NASA responded through official channels. Peter Badgley's office received the Interior announcement a few days before its release and immediately sent it up the hierarchy to NASA administrator James E. Webb, who discussed it with President Lyndon B. Johnson on September 20.[22] An announcement by the Secretary of the Interior required a response from a high level in NASA. Deputy Administrator Robert C. Seamans, Jr., wrote to Udall on September 22, presenting what appeared to be a decision from above about which agency would control the program:

> At this time, I believe that all affected agencies are fully aware of the potential value of such a program as EROS and of the enthusiasm with which the Department of the Interior has approached the applications of space technology to the hard problems of natural resources measurement. NASA is anxious to work closely and effectively with your Department in developing and testing this exciting application of technology prior to the decision to establish an operational system. Since many agencies are interested in pursuing similar efforts, and since there are a number of critical considerations involved (especially in the area of international relations), NASA is being assigned "lead agency" responsibility in *experimental* space applications for civil functions.[23]

NASA was allowed to define earth resource satellites as experimental, even though Interior claimed that some sensors were ready for operational use. Seamans's letter stressed the experimental nature of the project, cautioning that a "careful research and development effort must precede operational systems decisions if we are to assure that performance lives up to promise."[24] Thus, NASA kept control of earth resources satellites for an experimental phase of unspecified length.

The Department of the Interior continued to press NASA for a commitment to an operational satellite as soon as possible. After a meeting with Webb, Seamans and Jaffe, Under Secretary of the Interior Charles F. Luce wrote to Seamans: "Our staff people are quite optimistic that the state of the art has advanced to the point that we are ready for an operational system which we have called the EROS program. In staff discussions with Mr. Jaffe and his people we should be able to establish whether our optimism is justified."[25] On October 21, 1966, Luce transmitted to Seamans performance specifications titled "Operational Requirements for Global Resource Surveys by Earth Orbital satellites," calling for an operational system by the end of 1969.[26] These specifications (discussed in chapters 6 and 8) were reasonable; in fact, they closely resembled those for the experimental satellite that was eventually developed. When NASA provided no immediate reply, Pecora and Fischer continued to discuss an operational satellite to be launched in 1969. When Pecora led a discussion of earth resources satellites at a November 1966 meeting with officials from the Organization of American States, Jaffe (the NASA representative at the meeting) worried that Pecora was promising too much. Jaffe's memo on the meeting stated:

We agree on all points except whether anyone had the authority to hold out promise of an operational earth resources survey satellite as early as 1968 or 1969. Pecora took the position that since after the EROS public announcement no one specifically told them to stop, that they had every right to pursue and talk about EROS. We agreed that a policy clarification is required.[27]

The operational satellite that Pecora hoped to have in 1969 indeed turned out to be an elusive goal.

NASA's response to Interior's October data specifications was written in January and officially transmitted to Interior in April in the form of a letter from Seamans to Luce. It stressed the need for an experimental program, stating that "orderly development and flight qualification and testing of subsystems and sensors in an experimental development program permit assurances that the technology base exists on which to build the operational systems."[28] The comments attached to the letter described a system that could meet the requirements but saw the need for significant development of sensor, data storage, and data transmission technologies before they would be ready for a satellite.[29] Thus, NASA defined a long research and development program.

Effects of the EROS Announcement

The announcement of the EROS program was not a failure; it caused NASA to accelerate its earth resources research and development program. Although NASA officials were slow to respond to the requirements provided by Interior in October, the Department of the Interior announcement generated wide interest in the press, forcing the space agency to show at least steady progress. For example, an article in *Technology Week* in February 1967, headlined

"Earth Resources Satellite Far From Reality," pointed out that since NASA had not yet asked for any development funds, an experimental satellite could not be launched until the early 1970s.[30] Negative articles continued to appear for months. Jaffe explained to NASA Associate Administrator Homer Newell that an October 1967 article in *Aerospace Technology* "is obviously based on partial information resulting from the aftermath of the Department of Interior's EROS press release" and stated that relations with Interior "are extremely good at the moment."[31] A draft memo from Pecora to Luce in June also stressed good relations but explained:

The ungracious reception by Administrator Webb of the Secretary's announcement hindered full cooperative efforts for several months. The visit to Houston by the Secretary has helped measurably to alter Mr. Webb's views. Until some helpful statement by NASA appears, the technical press will continue to refer to NASA's earlier unfavorable position.[32]

By mid-1967, Interior and NASA had officially agreed on policy but not necessarily on how they construed it. An Interior offer to participate in funding was met with the response that it would be discussed further "if it proves desirable to proceed with an experimental satellite system."[33] In June, Pecora wrote with some frustration, "We continue to speak in terms of 'experimental system,' leaving the concept of 'operational system' for some distant time schedule."[34] Interior described the project as a joint research and development effort. NASA leaders, meanwhile, construed lead agency status as control and went ahead with the development of a satellite proposal without developing strong mechanisms for interagency participation in shaping the project.[35]

Despite these limitations, the group of enthusiasts at the Geological Survey felt that they had lost the battle but won the war. In particular, their announcement had resulted in the acceleration of activities at NASA. They had also established the EROS program, which continued to serve as an institutional base for satellite studies at Interior even when NASA was given responsibility for all satellite projects. However, they had also crystallized the opposing interests of some of the other agencies potentially involved by pushing for control of an operational earth resources satellite. The Department of Agriculture had been squeezed out, despite its equally valid interest in the program. This point became significant quickly, particularly when the time came to choose the sensors for the satellite.

The conflicting interests of the participating agencies were a fundamental problem visible from the beginning. NASA defined its interest as control of the experimental phase of the project and tended to see this phase as a long, incremental process of development. The user agencies wanted an operational satellite as soon as possible so that they could justify their interest by using the data in the routine execution of the missions they had been given by Congress. Even the Department of Defense was apparently involved, struggling behind the scenes to protect its interests by keeping the reconnaissance satellite program free from controversy.

6
Selecting the Sensors

The selection and design of the sensors for the Landsat satellite proved to be one of the most difficult and most politically constrained technical decisions of the project. Landsat incorporated much existing technology, pushing the state of the art only in the areas of sensors, tape recorders, and the data processing system. Therefore, negotiations among conflicting interest groups to define the technology played the largest role in those areas (see chapter 9 for a discussion of data processing). In the case of selection and development of sensors, NASA leaders had to negotiate a compromise between their interests and the different needs of the users within a framework of technical and political constraints and within limits set by the military and the budgetary process. This compromise provides a good case study of how different interest groups seek to define a technology for different purposes and how politics, user needs, and the development of technology interact. The negotiation of sensor definitions involved not only choices among available technology but also decisions that shaped the technical design of new sensor technology.

The Process of Selecting Sensors

NASA's role as a research and development agency and the agency's experience with sensors for planetary missions meant that space agency managers tended to see the development of sophisticated sensors as the first priority for the earth resources program. Engineers at the Johnson Space Center studied a variety of sensors from the beginning of the earth resources program in 1964, with a particular emphasis on developing a wide range of instruments to look at different wavelengths. However, NASA leaders assigned the preliminary design study for Landsat to the Goddard Space Flight

Center rather than to the Johnson Space Center. The Earth Resources Aircraft Program staff at Johnson was the only NASA team with extensive experience in the development of earth resources sensors but was busy at that time with Apollo Applications work. Goddard scientists and engineers had worked on meteorological and geophysical spacecraft but not earth resources. While they had as much interest in technical sophistication as the engineers at the Johnson Space Center, the Goddard engineers brought to Landsat a different approach to the users, expecting interested scientists to manage their own experiments.

The user agencies in turn had expectations of the sensors that did not always suit space agency interests. When the Department of the Interior's EROS team sent NASA specifications for an operational earth resources satellite, they expected that such a satellite would carry not the sophisticated sensors NASA was developing but a single, television-type sensor. In responding to those specifications and the political pressure behind them, NASA managers changed their emphasis from advanced sensors to those that might be ready for use in the next few years. They chose to develop the sensor the Department of the Interior wanted, but they also sought to counterbalance Interior's influence by including another sensor of interest to other groups both within NASA and at other agencies.

In the case of Landsat, NASA managers had more time than the planning process usually allowed to negotiate what instruments the first satellite should carry. Pressure from the Department of the Interior led NASA to start working toward a specific small earth resources satellite project in early 1967. The space agency studied three types of sensors for use on a simple satellite: photographic cameras, vidicons (television cameras), and scanners (which record a scene one line at a time). After completion of the research and preliminary design study in October of 1967, the next step would normally have been the final design and building phases of the project, but this was delayed because of difficulties in getting funds (discussed in chapter 7). Instead, for eleven months, through September 1968, Goddard earth resources personnel had funding to work only on developing better sensors. In October 1968 the satellite was finally approved, and full-scale design and development began.[1] This lengthy process of selecting sensors involved not just technical decisions but compromises that would affect how Landsat would be used, because different sensors and different versions of those sensors had advantages for different users.

The Return Beam Vidicon

The data specifications the EROS program sent to NASA in October 1966 were based on a particular sensor: a vidicon, a type of television camera, developed by RCA. The technology involved had grown out of the television camera RCA had developed for NASA's Ranger project, one of the scientific efforts that led up to Apollo by providing information about the moon. The Ranger spacecraft required the development of a fairly fine resolution sensor that could transmit images quickly because it took pictures while heading for—and crash-landing into—the moon.[2] Reluctant to disband its Ranger team at the close of the project, in 1965 RCA assigned them to develop a finer-resolution camera in hopes that uses for it could be found.[3] The Geological Survey had received early reports on this sensor in 1966, although RCA did not officially announce the new camera until October 1967.[4]

RCA's sensor, called a *return beam vidicon*, differed slightly from standard television cameras. As with a normal television camera, it used a shutter to snap a picture of the scene on the ground, briefly exposing a light-sensitive plate. In the case of the return beam vidicon this plate was made of semiconducting material. An electron beam then scanned the plate, and a signal was recorded when the beam was bounced back by an area of that sensitive plate that had been electrically charged by exposure to light. This system could achieve a resolution of 4500 to 6000 lines per picture. In other words, whatever the size of the scene that was focused on the sensitive plate, it could be read out as a picture made up of about 5000 separate scan lines, much finer resolution than the 525 lines in a U.S. commercial television picture.

The radio signal created by the sensor could be turned back into a picture on the receiving end. RCA printed the picture taken by the sensor by scanning photographic film with a laser beam, reproducing the modulations that had been detected on the sensitive plate. A laser beam was harder to guide than the electron beam traditionally used to reproduce television pictures onto photographic film (film can be exposed by electrons as well as by light), but RCA engineers believed that an electron beam would not be able to reproduce the fine resolution of the sensor.[5] Vidicon technology had reached a level of maturity where engineers had the information they needed to develop reliable systems, but the electron beam recorder was to a greater extent an unknown quantity.

The return beam vidicon obviously had potential for looking at earth resources from space, and the teams working on earth resources satellites at the Department of the Interior and at Goddard tended to emphasize it as the key sensor for Landsat. The vidicon fit the EROS specifications, and it was not a technology already in use by the Pentagon (see chapter 3). The Goddard team lacked experience with the wide range of earth resources sensors tested by the Earth Resources Aircraft Program at the Johnson Space Center, and they were familiar with the Ranger project. A November 1967 memo on the assignment of radio frequencies assumed (based on a Goddard study) that the first earth resources satellite would carry three television cameras with a resolution of 3300 lines per picture and that later satellites would use vidicons with improved resolution.[6] NASA gave RCA a small contract to develop the vidicon, and the Goddard earth resources satellite team showed its commitment to that sensor by pressuring RCA to speed up development. Oscar Weinstein of the Applications Experiment Branch at Goddard complained in a November 1967 memo that in eighteen months RCA had produced only one vidicon tube capable of the resolution claimed and that that tube was being used for demonstrations for potential customers rather than for research. Someone added a handwritten comment to the memo: "The battle continues to rage."[7]

To be sure, Goddard scientists and engineers had seen from the beginning that the vidicon would have some technical limitations, but they still felt that it was the best choice. The key problem was that any sort of television tube tended to distort the geometry of an image. The people who were going to use the image wanted it to be comparable to an accurate map, with straight lines for latitude and longitude. But the image produced by a television camera was like a map printed on a rubber sheet and then stretched and pushed out of shape, so the latitude and longitude lines were no longer straight. In other words, the geometry was no longer accurate. The heart of the television camera was a complex vacuum tube, which introduced geometric distortions from the glass of which it was made and further distorted the image because of the difficulties of exactly guiding the electron beam that read the picture off the sensitive plate. These distortions could be several percent of the image size, so that in a picture showing a scene 100 miles on a side, an object might appear 2 or 3 miles from its true position relative to other parts of the scene. Most of this effect did not change with time, so it could be measured before launch and corrected during data processing. To do so, however, would slow down data processing and increase the cost. Problems arose particularly for color pictures, where images taken by three different tubes in the three cameras equipped with different-colored filters had to be corrected to match each other.[8] Filters brought their own limitations, providing less accurate spectral information than other possible technologies. Nevertheless, Goddard plans for Landsat centered on the return beam vidicon technology because it seemed to promise fine resolution with the least technological risk.

The Multispectral Scanner

A different sensor technology, not closely related to a television camera, was already under development, but NASA had to be pushed to consider it seriously for earth resources. Instead of snapping a picture with a shutter, this sensor used a different type of system: A mirror scanned the scene on the ground directly, one line at a time (see figure 5). The mirror reflected the light onto a small array of photoelectric detectors, which produced an electric signal that depended on the intensity of the light reflected by the mirror from the scene onto the detector. Sensors of this type are called *scanners*. The scanning process introduced geometric distor-

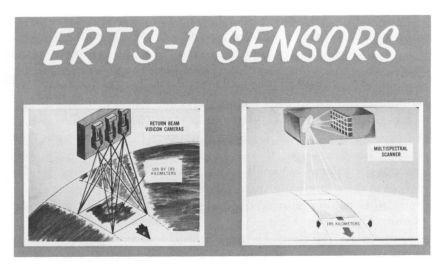

Figure 5
Landsat (originally called Earth Resources Technology Satellite) *1* sensors. (NASA, 1972)

tions just as the tube of the vidicon did, but the distortions caused by scanning later turned out to be more easily correctable than those of the vidicon. One-color scanner technology had originally been developed for use in the far-infrared band of the spectrum, where photographic film and television cameras do not work. For ordinary use it had the overwhelming disadvantage that any moving object in the scene would be blurred by the slow process of scanning one line at a time. But this was not a problem for taking pictures of earth from space. A special kind of scanner was developed to take color pictures: The mirror reflected light from the ground through a prism, which broke the beam up into different colors projected onto an array of photoelectric detectors, each of which measured the strength of part of the spectrum of the light coming from a single spot on the ground.

The technology that led to the particular scanner used on Landsat was developed at the Infrared Laboratory of the University of Michigan, which later became the Environmental Research Institute of Michigan. The laboratory had been founded in 1946 under the name Michigan Aeronautical Research Center, with a contract from the Department of Defense to investigate defenses against ballistic missiles. When this research was phased out in 1954–55, the laboratory received a large research contract to investigate reconnaissance and surveillance of battlefields. The work on this contract, known as Project Michigan, led to a wide range of research on infrared imaging. One part of this research involved the development of sensors for looking at the ground from aircraft, including multiple scanners that could be used to observe both visible and infrared wavelengths. These scanners had the disadvantage that they were extremely bulky, because a separate scanning mechanism was used for each wavelength band.[9]

Marvin R. Holter, later a vice president of the Environmental Research Institute of Michigan when it became an independent company, directed the technology developed for battlefield reconnaissance toward systems with both military and civilian uses. He pointed out to the military that information on vegetation and geography could be just as useful as locating people and tanks. This line of development suggested that the interactions between civilian and military technology involved not only scientists who recognized civilian implications in their military work but also the blurring of the distinction between civilian and military problems in an era of planning for total war. Holter's recommendation led to a research project in 1961-63, in which a multiple scanner system was flown over a fixed path every two hours for one day every two weeks for a year, to see what it could see. At the time, the technology was awkward: Four spectral bands were observed by flying two surplus military scanners with two spectral bands each in two separate aircraft, because the scanners were too large and heavy to fit in one of the available aircraft. The results of this study, declassified in 1964, provided good evidence of the usefulness of multispectral scanning.[10]

With support first from the Department of Defense and then from NASA, the Michigan team and a cooperating group at the Hughes Aircraft Company then invented a true multispectral scanner. This scanner would make color pictures without using a separate scanning system for each spectral band. Virginia Norwood at Hughes developed the idea of using a prism to separate different spectral bands from the light provided by one scan of the mirror. This instrument was originally classified, but it had to be declassified when a patent search revealed that the idea had already been patented and that the earlier patent was not classified.[11] The multispectral scanner provided good color pictures and also had another advantage: The prism and detectors could be arranged to measure the light in a very narrow portion of the spectrum (fine spectral resolution), and light outside that band would not be measured at all. This was an improvement over a sensor that selected the spectral band with a filter in that it cut off unwanted parts of the spectrum more sharply. These advantages were sufficient to persuade NASA managers that the multispectral scanner deserved further research. In 1967 NASA issued a contract to Hughes to further the development of the multispectral scanner, which the NASA procurement justification described as "an advancement in the state-of-the-art in the field of Electro-Optical Systems [that] cannot be duplicated by another source."[12]

Gradually, some scientists doing research under contract from NASA realized that the multispectral scanner deserved to be considered for an early satellite experiment because it might be more useful than a return beam vidicon for agricultural research. Geologists had limited interest in the new sensor because they expected it to have poor geometric accuracy. But its simpler production of color pictures and more precise spectral bands was of great potential value in agricultural studies, where spectral information provided the key to differentiating crops. The infrared bands that the scanner could provide were also seen as particularly useful for agriculture because they could give information on crop health.

The perceived advantages of the multispectral scanner resulted in a growing interest in using it for earth resources, particularly at the Department of Agriculture. As late as early 1968, Archibald B. Park, the head of Department of Agriculture research on remote sensing satellites, and Oscar Weinstein, a sensor engineer at Goddard, proposed the RCA return beam vidicon rather than the scanner as an earth resources experiment in the Nimbus weather satellite program.[13] As time went on, however, Park switched his emphasis, pushing NASA to include a multispectral scanner on the earth resources satellite being planned, despite the reluctance of Goddard engineers (who feared that the technology might need considerable further development).[14]

Persuading NASA to include the multispectral scanner on Landsat required compromises both within NASA and among agencies. The scientists at the Geological Survey, having prepared to work with the vidicon since 1966 and believing it better suited to geology, had little interest in the scanner. In October 1968 a memo written by Goddard scientist William Nordberg outlined Goddard's position: he stressed the importance of developing a multispectral scanner, however, he assumed that it would not be flown on Landsat but perhaps on *Nimbus F*.[15] However, higher levels at NASA had different interests than the people at Goddard. In particular, NASA leaders wanted to balance the Department of Interior's interest in taking early control of earth resources satellites with participation by other agencies. Therefore, just as the Department of the Interior had been able to pressure NASA to proceed with a small satellite, the Department of Agriculture was able to persuade NASA to add the scanner to Landsat. In February 1969 the interagency Earth Resources Survey Program Review Committee approved Landsat specifications that included both the vidicon and the scanner.[16]

Photographic Sensors

A third option, the use of photographic sensors, was also considered during studies conducted at Goddard in 1967 and 1968 on what type of earth resources satellite to fly. Geologists, and particularly the National Research Council's Committee on Space Programs for Earth Observations (which had been formed to advise the Geological Survey on EROS), called for a satellite that would take fine-resolution photographs and return the film to earth. Such a film-return satellite could provide finer-resolution images with less distortion than either a vidicon or scanner and would, therefore, be particularly useful for mapping.[17] Such technology had been used for many years by the Department of Defense (see chapter 3) but turned out to be unavailable for civilian use.

Some discussion took place between NASA and Department of Defense officials on the possibility of declassifying military technology pertaining to film-return satellites and infrared scanners. The civilian side made the argument "The country must not accept arbitrary classification, but must constantly assess return in security versus economic costs."[18] A paper arguing for declassification of an infrared scanner also pointed out that the restrictions of security classification tended to slow further progress on a technology, thus possibly decreasing, rather than increasing, the U.S. advantage.[19] An industry journal claimed in March 1968 that "an important high level debate is raging" between NASA and the Department of Defense, explaining that "some sources interpret Pentagon resistance to allowing NASA use of high resolution surveillance systems as a step to monopolize jurisdiction of photographic-infrared earth observation systems."[20] This debate resulted in the declassification of some infrared systems for use from aircraft, according to a complex formula. In order for a system to be unclassified, its resolution, temperature sensitivity, and the ability to image a large area quickly (combined into a variable called *order-of-merit*) could not exceed a stated value.[21]

However, NASA more often tended to avoid battles with the Department of Defense. Despite its civilian image, NASA often worked closely with the Pentagon; assistant to the administrator for special projects David Williamson wrote in an outline for a telephone conference, "I have worked hard to make NASA an effective part of the national security establishment without compromising the space act."[22] Civilian earth resources satellites generated a great deal of opposition from the national security community. A memo written in May by a retired air force general, who was a consultant to the NASA Administrator, explained that "in the Intelligence Community, at the working levels at least, there is the traditional, understandable feeling that almost any earth sensing from satellites by NASA, with their inevitable public disclosure, will be harmful, and perhaps fatal, to classified sensing for national security purposes."[23] NASA and the Department of Defense finally agreed that NASA could fly vidicons and scanners with coarse resolution, but not film-return satellites.

An interest in photographic cameras persisted, however, because many potential users of Landsat data had experience with aerial photography and wanted something similar from a satellite. For a time, NASA's long-range plans called for *Landsat A* and *B* (to become *Landsat 1* and *2* after launch) to be followed by two film-return satellites, *Landsat C* and *D*.[24] This scheme finally died when the Department of Defense told NASA to stop and threatened to convince the Bureau of the Budget to cut Landsat funding. At that point, the scientists who had been pressing for a film-return satellite "decided that it would not be in the best interest of the country as a civilian experiment."[25] In fact, the Bureau of the Budget refused to allow NASA even to list mapping as an objective of Landsat because the Department of Defense already utilized classified satellites for mapping. For NASA to do the same thing would be a duplication of effort.[26]

Specifications

Selecting types of sensors was only one part of the process of negotiation; the specification of those sensors also involved critical disagreements about technical goals. In particular, NASA still had to determine how fine a resolution Landsat users would need to achieve their goals and whether this resolution could be achieved within technical and military restrictions. The users, familiar with fine-resolution aerial photographs, tended to specify that they needed a similar resolution from satellite data in order to gain substantial benefits. A September 1968 letter from Archibald B. Park at the Department of Agriculture to Leonard Jaffe at NASA stated: "It is obvious that the real payoff for agriculture requires resolutions better than 10 meters."[27] A Westinghouse cost-benefit study commissioned by the Department of the Interior concluded that benefits would be doubled if resolution was improved from 100 feet to 30 feet and halved if resolution slipped to 300 feet. For the Department of the Interior alone, the study predicted an "effectivity increase" or benefit of $218 million at 10-foot resolution, $160 million at 30-foot resolution, $112 million at 100-foot resolution, and $40 million at 300-foot resolution.[28]

NASA managers believed that fine resolution would cause controversy with the national security establishment and be prohibitively expensive, so the space agency set out to convince the users that coarser-resolution data would be most efficient. Fine-resolution data would be extremely expensive to process simply because of the huge amount of data produced—if the resolution of a picture were improved by a factor of three, the amount of data would increase by a factor of nine. This could quickly become a problem for a satellite taking pictures of the entire world. On the other hand, repetitive coverage and spectral data could partially make up for lack of resolution by revealing differences, such as the color of fields, that did not show up even in fine-resolution black-and-white images. The user agencies fairly quickly accepted NASA's arguments about the disadvantages of extremely fine resolution, but there were still complaints about the resolution NASA decided to provide. The Department of the Interior's September 1968 statement of requirements concluded:

While our studies suggest a gross benefit increase with improvement of resolution from 100 to 30 feet, such an improvement will increase cost of data processing approximately 10 times. Therefore, our general requirement for images having a spacial [sic] resolution of 100 to 200 feet . . . remains unchanged.[29]

Many users saw the 100- to 200-feet requirement as a minimum; some scientists still believed that finer resolution was necessary, and there were protests from the user agencies in 1970 when it became clear that the vidicon would give a resolution of only about 300 feet.[30] By 1970, however, most users had agreed that resolution was not as important as they had earlier thought. Despite this, the Bureau of the the Budget later claimed that Landsat would not be useful because it did not meet the original requirements of the users (see chapter 7).

In addition to resolution, NASA also had to reach an agreement with the user agencies about the spectral bands, or colors, that the sensors would observe. The vidicon and the scanner provided data in only three and four spectral bands, respectively. Different sets of spectral bands were useful for different disciplines, so NASA had to negotiate a compromise among scientists in different disciplines both within NASA and in the user agencies. This disagreement came to a head in a May 1969 meeting of NASA scientists working with scanners at which "there was general agreement on the types of scanners needed but serious disagreement on the wavelength intervals to be used."[31] The spectral band selected by the Goddard study for the scanner on Landsat followed the recommendations of the Department of Agriculture and had been agreed to by the Geological Survey, which had little interest in that sensor. The scientists working with aircraft scanners at the Johnson Space Center disagreed, claiming that not enough was known to select satisfactory spectral bands for a scanner on a satellite.[32] In that case the primary issue revolved around NASA scientists, who wanted to do more research. On the other sensor, the vidicon, the Department of the Interior initially proposed two of the three spectral bands. NASA discussed alternatives with the user agencies and, at the request of the Department of Agriculture, added a third channel in the red region of the spectrum in addition to the blue-green and near-infrared requested by Interior.[33]

Even these compromises did not fully meet the needs of agricultural users. Representatives from the Department of Agriculture had recommended spectral bands for the multispectral scanner that would detect the peaks and valleys of the average spectrum of green vegetation.[34] The spectral bands finally selected for Landsat were too wide to show individual features in the spectrum of particular crops, which was what would have been most useful to the agricultural scientists. Goddard engineers argued that narrower spectral bands were not possible on Landsat, because the interrelationship of spectral resolution, spatial resolution, and the speed at which the sensor scanned the scene on the ground made sensor technology providing narrow spectral bands extremely difficult to achieve.[35] Studies also had shown that a spectral band in the middle of the infrared (1.5–1.8 microns) was particularly useful for differentiating vegetation, but that band was also technically difficult.[36]

When the first satellite was launched in 1972 (see figure 6), the two sensors provided some surprises, but the technology was generally satisfactory and established a base for further developments. The vidicon failed after a few days, forcing the scientists to learn to use the data from the scanner. Scientists at the Goddard Space Flight Center, the Geological Survey, and the contractors developed geometric corrections for the scanner data, which allowed more geometric accuracy than the Department of Interior scientists had thought would be possible. The second Landsat satellite carried sensors identical to the first satellite, whereas the

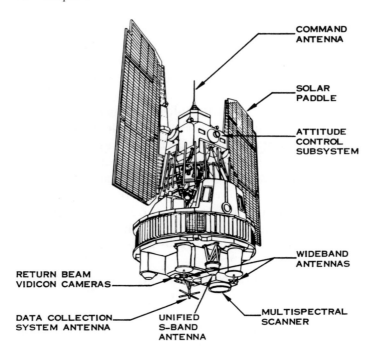

Figure 6
Landsat configuration. (NASA, 1972)

third included some minor modifications. A far-infrared spectral band, which would pick up the radiation emitted (not just reflected) by objects, was added to the scanner but functioned poorly. Since the vidicon had proven to be less useful than the scanner as a source of color pictures, the three-camera system was replaced by two (black-and-white) cameras with improved resolution—100 feet instead of 300.[37] Studies for a more advanced earth resources satellite called Earth Observations Satellite had called for the development of a fine-resolution sensor that would be pointed at particular areas of interest rather than covering the whole earth and the inclusion of additional spectral bands for various disciplines.[38] When this project was not approved, NASA developed an improved scanner called the *thematic mapper*, with finer resolution and more spectral bands, for the fourth and fifth Landsat missions.

The process of selecting sensors showed the degree to which technical decisions were also negotiations about the usefulness of the satellite system, since different users preferred different systems. The selection and design of sensors was not so controversial as the development of the data processing system (discussed in chapter 9), but it does illustrate how NASA made compromises among the needs of the various agencies within the context of NASA's interest and the technical limitations seen by NASA engineers. By making the satellite a compromise, rather than ideal for any one agency, NASA strengthened its control of the program, weakening the argument for any one user agency to take over management of the satellite program.[39] Thus, the process of defining the satellite involved not only choices of technology but assumptions about and approaches to those choices.

7
The Battle for Funding

NASA and the user agencies not only had to agree on sensors for the Landsat satellite, they also had to obtain the money to design and build it. During the late 1960s NASA had great difficulty getting new projects onto the federal budget, and Landsat met a particularly hostile reception from the Bureau of the Budget.[1] Opposition from the Budget Bureau forced NASA and the user agencies to work together more effectively than they had done previously. It also added to the tension between research interests at NASA and operational interests at the user level. The Bureau of the Budget complicated the situation by imposing its own definition of the Landsat project. The Budget Bureau insisted on what NASA managers saw as contradictory requirements: agreeing to fund only a minimal experimental project while requiring that interested agencies defend earth resources satellites with cost-benefit calculations for an operational system. The Bureau's approach grew out of its analysts' attitudes towards the space program in general and Landsat in particular and also out of broader changes in budgetary policy and the budgetary process in the late 1960s and early 1970s. This study deals with the role of the Budget Bureau only in its interaction with NASA and the Landsat user agencies.[2]

The Budgetary Process

Different projects hit opposition in different stages of the budgetary process; for Landsat, difficulties arose primarily during negotiation to determine the budget the president would send to Congress. That process began about two years before the year the money would be spent. The Bureau of the Budget had responsibility for preparing the budget, but the process started when each agency drew up its own proposed budget, within guidelines set by the Budget Bureau. The Bureau of the Budget then integrated these agency budgets into a single federal budget, often making substantial changes.

The president's budget was often not the final word. The president presented that federal budget to Congress about six months before the beginning of the fiscal year during which the money would be spent.[3] Congress, of course, could alter the budget before passing it. NASA sometimes found more support in Congress than in the White House, but as political appointees, NASA administrators were obliged to defend the president's budget even when Congress wanted to change it to something closer to what NASA had originally requested. Once Congress passed a budget, it then had to be signed by the president before it took effect. The Bureau of the Budget normally distributed money according to the budget that the president had signed but occasionally chose for policy reasons to withhold funds that had been appropriated.

One major change took place in the general budgetary process during the period of controversy over Landsat funding. In 1970 President Nixon reorganized the Bureau of the Budget and renamed it the Office of Management and Budget. That reorganization increased the management function of the budgetary agency: Increasingly, it not only prepared the budget but also controlled the spending of money after it had been appropriated. NASA leaders disliked this change. Most obviously, it reduced the extent to which agencies could control their own funds, but it also changed the role of the budget agency to that of "a much more political tool rather than an honest broker."[4] The new Office of Management and Budget had an additional level of political appointees, and many of these new political appointees saw their agency as a tool for fulfilling their conservative vision of the proper role of the federal government.

Obtaining Funding

NASA managers knew from the start that it would not be easy to obtain approval and funding for Landsat. The former director of NASA's Office of Space Science (which also included applications from 1963 to 1972), Homer Newell, pointed out in a history of that office that NASA generally had a harder time getting funding for applications than for science. Science was supported for its own sake, but applications had to be proven to be "practical" before a program was initiated. Moreover, the administration of Richard M. Nixon, says Newell, "was not inclined to encourage investment in expensive new systems—even if they were better— when the old systems were adequate."[5] Initiating a new project was particularly difficult, because approval of a project as a line item in the budget represented a fairly strong commitment to continued funding. NASA might have more easily gained approval to fly earth resources sensors on approved missions in the human spaceflight program or on experimental satellite projects like the Nimbus weather satellite series or the Applications Technology Satellites (which carried mostly communications experiments). But NASA wanted to initiate a new satellite project for earth resources (see chapter 4).

Initially, funding for research into remote sensing of earth resources came from other NASA programs. Peter Badgley started the earth resources program in 1966 with funds budgeted for unspecified exploratory research and money set aside to define experiments for Apollo Applications missions. In fiscal year 1967 earth resources became a separate program funded in the manned space budget for $4.2 million. This provided research funding but did not constitute approval of a satellite project. Additional funds came from other parts of the budget, particularly Apollo Applications, to investigate sensors as potential experiments for those projects. Table 1 illustrates authorized funding for earth resources projects and total levels of funding for some items related to earth resources. Table 2 shows funding at the Department of the Interior, the only user agency with a separate budget item for earth resources satellite spending.

In order to design and build a satellite, NASA had to have specific funding for the project, not just general research funds, and such project funding proved difficult to obtain. The space agency's request for $4 million for an earth resources satellite in fiscal year 1968 was initially reduced by the Bureau of the Budget to $1.5 million and then rejected entirely. This cut apparently resulted from the conflict with the Department of Defense discussed earlier.[6] After an appeal the Budget Bureau allowed some funding for fiscal year 1968 but at a much lower level that NASA managers believed necessary. For example, one of the major activities of this stage of the project was research from aircraft to investigate what sensors would give what information (see chapter 5). The 1968 budget was so low that NASA could provide only 25 percent of the funds

Table 1
NASA Programmed Funding for Earth Resources (in millions of dollars)

	Satellite program	Aircraft program	Manned space science
1967			4.2
1968			0.3
1969	2.3	8.9	
1970	15.0	11.0	
1971	55.75	10.99	
1972	52.08	12.35	
1973	32.6	13.0	
1974	17.4	16.6	
1975	17.7	7.7	
1976	18.1	5.5	
1977	21.2	12.75	
1978	61.96	9.1	

Source: *NASA Historical Data Book*, vol. 1 and 3

Table 2
Department of the Interior Funding for Earth Resources Satellite Systems (millions of dollars)

FY	From NASA	Redirected	Appropriated
1964	0.1		
1965	0.3		
1966	2.35		
1967	2.18		
1968	3.6	0.173	
1969	1.832		0.2
1970	1.94		1.1
1971	1.866		1.921
1972	1.035		5.75
1973	2.835		10.043
1974	0.057		8.963
1975			8.284
1976			10.395
Transition quarter			2.611
1977			9.545
1978			9.735

requested by the user agencies for research with sensors carried in aircraft.[7]

In the fall of 1967 NASA and the Department of the Interior again proposed an earth resources satellite project, this time for the fiscal year 1969 budget. At the time the satellite was called Earth Resources Technology Satellite; its name was changed to Landsat in 1972 (it is referred to as Landsat throughout this study to reduce confusion).[8] Interior requested $3 million, while NASA asked for $25 million, of which $15 million would go for component and spacecraft development and the rest for research.[9] The Bureau of the Budget turned down both requests.

In the end NASA obtained approval for Landsat that year—but only after a long fight. In late 1967 NASA Administrator Webb met with the director of the Bureau of the Budget appealing for restoration of four major cuts, including $17 million for Landsat. The Bureau of the Budget was not sympathetic to NASA's protest. A memo from D. E. Crabill of the Economics, Science, and Technology Division of the Bureau of the Budget questioned "to what extent Mr. Webb is appealing this item from personal conviction and to what extent he feels coerced to do so to demonstrate good faith with the Vice President and the Secretaries of Interior and Agriculture." The memo went on to strongly recommend against any significant restoration both on the technical grounds that Landsat was "oversold and premature" and on the political grounds that "the two principal user agencies have already accepted our position."[10] In a meeting between the space agency and the Budget Bureau to discuss the issue, NASA's Acting Deputy Administrator Homer Newell argued that a satellite proposal should be approved despite uncertainties about the benefits an operational earth resources satellite system would bring. He maintained that actual satellite data were necessary for the research that would prove whether an operational earth resources satellite system would be worthwhile. He also threatened that Webb would discuss the need for Landsat funding during congressional hearings whether the Budget Bureau approved it or not.[11] When the Budget Bureau was not swayed, Administrator James Webb finally appealed all the way to President Johnson for approval of the satellite project and won a partial victory: a budget of $12.2 million for earth resources, of which $10.2 million was slated for research using aircraft to test the concept and $2 million for studies and development of sensors for a satellite.[12]

Congress was more favorable to the project than the executive branch. The chairman of the House Subcommittee on Space Science and Applications, Joseph E. Karth (D-Minn.), had become enthusiastic about the value of an earth resources satellite and wanted NASA to speed up the program. Disgusted by the small request the administration was making for Landsat, Karth criticized both the Bureau of the Budget and NASA from a point of view he expressed clearly a few years later: "In looking at the history of ERS [Earth Resources Surveys], I come to the inescapable conclusion that there is a preponderance of evidence of footdragging, setting up of strawmen, and the assignment of unique and unusual and, I might say, ridiculous yardsticks and so on and so forth."[13]

Karth did not succeed in raising the appropriation for Landsat, but Congress did rearrange the budget to speed progress toward actually building a satellite. While the total budget for earth resources for fiscal year 1969 remained at $12.2 million, Congress changed the distribution of the funds: $7.7 million was earmarked for research using sensors carried on aircraft, and $4.5 million for satellite studies and sensor development.[14] NASA could finally begin the actual development of the Landsat satellites, spending $1.2 million of the appropriation on competitive studies by industry of preliminary designs for an earth resources satellite. Most of the remainder of the $4.5 million went for sensor development already underway.[15]

The Earth Resources Observation Satellite (EROS) program at the Department of the Interior had less success than NASA for fiscal year 1969. Interior had requested $3 million for EROS, primarily for a data center to process and distribute Landsat data. This facility had to be funded immediately so that it would be operational in time for the first satellite mission. Despite this argument, the Bureau of the Budget cut the EROS request to $200,000, although this represented some improvement over the previous year when the bureau eliminated it from the final budget.[16]

NASA had more success with its fiscal year 1970 budget than it had had the previous year. Some Bureau of the Budget staffers were still very critical; one internal Budget Bureau memo in June 1968 suggested that the Landsat satellite proposal should be eliminated and replaced with an earth resources aircraft program.[17] But NASA provided the new studies of the usefulness of Landsat that the Bureau of the Budget had requested and convinced critical Budget Bureau analysts that the project was worthwhile. A November 1968 staff paper from the Engineering, Sciences, and Technology Division of the Bureau of the Budget recommended $14.1 million in funding for Landsat because "we believe that the case for potential economic benefits has been established."[18] The leaders of the Budget Bureau accepted that recommendation; the space agency received $14.1 million for Landsat and another $11 million for the earth resources aircraft program, enough to let a contract to design and begin building the satellite.

Unlike NASA, Interior did not have more success that year than previously. Funding for Landsat data processing at Interior had originally been cut because of doubts about whether NASA would get funding to proceed with the satellite program. When NASA's Landsat funding was approved, Secretary of the Interior Stewart Udall offered to make other cuts to keep the total departmental budget below the ceiling set by the Bureau of the Budget while restoring funding for Landsat data processing at the Department of the Interior. The budget analysts most immediately concerned, in the National Resources Program division of the Bureau of the Budget, recommended the restoration, but it was not approved.[19] The president's final budget allotted to EROS only a small fraction of the funds requested. Congress assisted the project, increasing the authorization to $4.1 million. The Bureau of the Budget, however, impounded $3 million of that, leaving EROS with only $1.1 million.[20] At that time the law did not specifically prohibit the Bureau of the Budget from refusing actually to spend money that had been appropriated by Congress. As a result of continuing budget problems, the EROS Data Center was reduced in scope and delayed by two years. In the long term these delays resulted in serious problems with the distribution of high-quality data to experimenters during the first few years of Landsat operations (see chapter 11).

During congressional hearings on the 1970 NASA budget Congressman Karth battled the remaining opposition in the executive branch. When Homer Newell's replacement as head of NASA's Office of Space Science and Applications, John E. Naugle, explained that the project had been delayed by studies requested by the Budget Bureau and the president, Karth "expressed regret that there are others who, unlike this Committee, 'do not accept your word as scientists...and require you to go through costly studies.'"[21] When Director William Pecora testified to express the Geological Survey's support for NASA's fiscal year 1970 budget request:

Mr. Karth asked which agency had been the least cooperative. Dr. Pecora compared the inquiry to the question of quality among beers: none are bad; it is just that some are better than others. This was not acceptable, and, after much verbal maneuvering, the initials that emerged were BoB [Bureau of the Budget].[22]

Naugle and Pecora had to be loyal to the executive branch, but they also were reluctant to criticize the Bureau of the Budget when it was beginning to be more supportive of Landsat.

However, the support NASA had won for Landsat at the Bureau of the Budget tended to erode after Richard Nixon took office in 1969. Even with some funds already committed to building a satellite, the budget battle was fought all over again when NASA and Interior submitted their fiscal year 1971 requests. In its revisions of the agency budgets in the fall of 1969, the Budget Bureau eliminated earth resources entirely from the Department of the Interior's request and cut NASA's budget for Landsat from $41.5 million to $10 million by delaying any launch of a satellite until after 1972. The agencies were outraged. Interior argued that the program had strong public support and its elimination would be an "abrupt change in Administration policy" that "cannot help but become a major embarrassment to the Administration."[23]

The Landsat enthusiasts had a new weapon; President Nixon had publically committed the country to an earth resources satellite program in a speech before the United Nations in September 1969.[24] But the Budget Bureau still argued against funding Landsat, claiming that the Landsat sensors did not meet the specifications of the users.[25] The new NASA Administrator, Thomas O. Paine, went to President Nixon, disputed this claim, and persuaded the president to restore NASA's request. Interior, however, was left with only $1.92 million of its $11.6 million request.[26]

The fiscal year 1971 budget decisions effectively committed the Nixon administration to the *first* Landsat satellite, and the Bureau of the Budget honored this commitment by providing the minimum necessary funding. Almost identical problems arose, however, in getting funding for the subsequent satellites in the Landsat series.[27] As early as the spring of 1971 the Office of Management and Budget was trying to cancel the backup satellite, *Landsat B*. NASA argued successfully that a backup was essential, and after the success of *Landsat 1* in 1972, the space agency found enough support to proceed with *Landsat B*, launched as *Landsat 2* in 1975.[28] Despite continued controversy over funding, NASA was able to launch *Landsat 3* in 1978 with improved sensors, *Landsat 4* in 1982 with the addition of a new, finer-resolution sensor, and the virtually identical *Landsat 5* in 1984.

Reasons for Opposition

Opposition to Landsat at the Bureau of the Budget had complex causes. Dislike of Landsat grew both out of presidential agendas (particularly under Nixon) and out of the opinions and experience of budget analysts. Unlike the situation with major space projects, in the case of Landsat the contributions of the bureaucracy to funding decisions played a larger role than presidential policy because Landsat was too small a project to get much presidential attention. Of course, the general climate established by the president played a role, but a full study of that effect would require a case study of a project that gained a higher place on the agenda. Most interesting from the history of technology point of view is the effect of Bureau of the Budget opposition on technological decisions. Simple lack of funds affected technological choices, but, more importantly, satellite design was affected by the arguments against Landsat used by budget analysts and the specific types of support that the Budget Bureau required NASA to provide to prove that Landsat was worthwhile.

The Bureau of the Budget opposed an earth resources satellite project for at least four reasons. First was the desire to decrease the federal budget in general and NASA's budget in particular, which meant that all new projects came under careful scrutiny.[29] Second, personnel at the Bureau had experience with the many problems that plagued new projects, particularly those intended for widespread practical use. The Bureau of the Budget doubted, with good reason, that NASA officials had enough experience with such projects to understand fully how difficult it was to transform a new idea into a usable, profitable system.[30] Third, the Bureau of the Budget staff, a number of whom were employees of the military or the CIA detailed to the Bureau, worried about the risks of calling attention to military reconnaissance satellites, which they considered vital to national security.[31] Fourth, the budget watchers still had basic questions about whether satellites could perform earth surveys as well as aircraft. The fourth reason became the public focus of the debate.

The aircraft-versus-satellites debate involved a variety of issues and hidden agendas. The champion of aircraft surveys was Amrom

H. Katz, an influential aerial reconnaissance specialist who had worked in reconnaissance during and after World War II and had been employed at RAND Corporation since 1954. Katz participated in a National Academy of Sciences summer study of applications satellites in 1967, where he suggested that a program of photography from high-altitude aircraft would be cheaper than a satellite program. It would also raise fewer political problems because the aircraft would fly over only those countries that requested photographs.[32] Katz pressed his argument in an article in *Astronautics and Aeronautics* in June 1969 and in a public confrontation with NASA's Leonard Jaffe at an American Institute of Astronautics and Aeronautics meeting on earth resources information systems in early 1970. NASA officials used the meeting to make potential users aware of what the satellite would do and to generally promote Landsat, which was becoming politically useful as an example of the benefits of the space program. In an August 1970 article, Katz complained about the NASA presentation at the meeting:

> The Harlem Globetrotters get around, but the wandering road show featuring discussion of earth-resource surveys must run a close second.... The issues, and the real problems, were hidden, like strawberries in a bowl of viscous yogurt, cliché-ridden, and made up of "in" words and religious references to environment, priorities, 3rd world needs, population, resources, etc.[33]

Katz used a variety of arguments to support his position. He attacked NASA's Landsat design for not meeting the resolution requirements specified by the 1967 summer study. He supported his argument with impressive figures comparing costs of aircraft with those of film-return satellites, a design NASA was not considering. Compared with satellites, aircraft could provide finer resolution and lower cost for short-lived missions; however, NASA managers maintained that a satellite carrying a television-type sensor that continued to take pictures for a year or more would cost less that aircraft missions. Satellites could also cover parts of the world where aircraft would not be allowed to go and would provide regular repetitive coverage to show changes. Unconvinced by NASA's position, Katz continued to argue for the superiority of aircraft at least through 1976, long after the launch of *Landsat 1*.[34]

The debate between Katz and supporters of earth resources satellites provides an extreme example of debates over the definition of a technological system. Katz argued that aircraft and satellites were competing systems to produce the same data and therefore could be compared on the basis of development and operating costs. However, the systems whose cost Katz calculated produced data of different types, extent, and frequency than those that NASA planned would result from Landsat. Katz's definition of a satellite system was motivated both by his previous experience and by his interests in defending classified systems.[35] The resulting debate rarely addressed the differences in product and therefore tended to make little progress. Since the system Katz proposed was not tested, it was impossible to compare its usefulness to that of Landsat. Yet, the continuing use of aerial survey for a wide range of needs suggests that a large-scale earth resources aircraft system would have complemented rather than competed with earth resources satellites.

Katz's attack had considerable impact. Most importantly, he had friends at the Bureau of the Budget who listened to his arguments. In fact, William Fischer of the Geological Survey said that many people involved with Landsat believed that Katz was acting as a spokesman for the people at the Department of Defense who were worried that a relationship might be perceived between Landsat and reconnaissance satellites.[36] Although these suspicions could not be proven, many of the issues raised by Katz appeared in the controversy over Landsat funding. As early as the summer of 1968 a Budget Bureau analyst argued that an aircraft program would be far less expensive than satellites.[37] The Bureau of the Budget's argument for cutting Landsat development from the fiscal year 1971 budget also seemed to show Katz's influence. The bureau's criticisms included the following:

1. The expected spatial resolution of the sensors does not conform to the requirements initially specified by the user community.
2. NASA has stated it will fly ERTS [Landsat] even if the sensors cannot meet specifications by the users.[38]

In a protest to the Budget Bureau, NASA officials pointed out that the users had learned that they could use data of coarser resolution than they had indicated in their initial predictions of operational requirements.[39] A NASA outline for a presentation a few weeks later, in December 1969, argued that aircraft and spacecraft were useful for obtaining different types of data and that the resolution expected from Landsat had been proven useful by a photography experiment (S065) simulating Landsat that was performed on *Apollo 9*.[40] The Budget Bureau argued from a classified study that stressed the need for fine resolution, but NASA replied that the study provided neither multispectral nor repetitive data, both of which would help compensate for coarse resolution.[41]

Katz's arguments that Landsat was not the most economical approach to earth resources survey and the opinions at the Bureau of the Budget that reflected those arguments fed into what NASA managers saw as an insidious demand on Landsat: that the agency should first prove that the benefits of the project would exceed the costs. A requirement for cost-benefit studies later became a common political attack on NASA, but its early use in the Landsat debate appears to have stemmed from security issues. The reconnaissance satellite community considered a civilian earth resources satellite to be a risk to national security. Therefore, they argued, NASA should prove that the benefits outweighed the risks. The risk to national security might be impossible to quantify, but studies could at least compare the value of the satellite data with the price of the project. The Department of Defense, particularly through air force contracts with the RAND Corporation, had developed the necessary techniques of cost-benefit analysis.

Probably because of Department of Defense pressure, the Bureau of the Budget required NASA to justify Landsat on a cost-benefit basis.[42] Interior had contracted with Westinghouse for the first cost-benefit study of Landsat in 1967, which estimated benefits (some of which are discussed in chapter 11) to the Department of the Interior alone at $80 million per year and known costs of a satellite project at $32 million per year.[43] The budget agency demanded additional studies from NASA, addressing not only what benefits would accrue but also "what actions would be necessary to reap the benefits from space-aquired data and what would be the political implications of these actions."[44]

After NASA and the user agencies had answered the questions about the resolution that Landsat would provide, the Nixon administration criticized the project primarily by requiring more cost-benefit analyses. Charles Robinove of the Department of the Interior's EROS program reported that one budget examiner explained that a scientific program could be funded if supported by the scientific community, "but if you're saying this program is going to be useful to people, will provide benefits and so on, then before you get the money for it you're going to have to prove there's a favorable benefit-cost ratio."⁴⁵ A letter from Budget Director Robert P. Mayo to Secretary of the Interior Walter J. Hickel in 1970 said:

> The Administration does not want to move beyond the experimental phase of this program until we are confident that the benefits of the program are more than the expected costs. Moreover, the evidence is unclear as to the technological capability of ERTS [Landsat] to provide pictures with sufficient resolution to give the payoff claimed in some studies that have been undertaken thus far. Until we have greater assurances on these points the financial commitment should remain at a minimum level.⁴⁶

Certainly the Bureau of the Budget had the right to ask for evidence that Landsat would in fact be useful. The Budget Bureau correctly perceived that NASA did not always seek to serve user interests. For example, a note by NASA Administrator James C. Fletcher in response to a 1974 study by the National Academy of Sciences supporting Landsat read, "I wish we could fight on science grounds instead of 'users.'"⁴⁷ However, NASA had to accept the cost-benefit challenge. The space agency funded cost-benefit studies by the Earth Satellite Corporation and the Environmental Research Institute of Michigan in 1972, and others in later years, all of which showed impressive benefits.⁴⁸

On the other hand, NASA managers argued that predicting with accuracy the benefits of a future technology was an impossible task. The experimental project could not be expected to pay for itself, so it was necessary to predict what the operational system would be before calculating costs and benefits. Landsat program officials believed that the data from Landsat would provide information never before available that would be beneficial to a variety of users, but potential benefits resulting from the data could not be known with any reliability until the system was operating. Landsat's staunch supporter Congressman Karth said of the required cost-benefit studies: "I know of no precedent for such a requirement anywhere else in the space program . . . . One shudders to think what the results of a cost-effectiveness analysis would have been for the Apollo program."⁴⁹ Because the Bureau of the Budget expressed reluctance to spend more money even for more benefits, NASA had to try to argue that Landsat could save the government money by providing data previously obtained by more costly aircraft and ground surveys. Enthusiasts for Landsat at NASA found this particularly frustrating because it ignored all the unique advantages of satellites, particularly the possible uses of data that could not be obtained by any other means.

The emphasis on cost-benefit analysis led the Budget Bureau to place specific requirements on Landsat. The project was permitted to go forward, but strictly as an experiment. The budget director wrote: "We must emphasize that this program is an experimental effort. We do not believe that it is prudent to invest substantial resources at this time in preparing for an operational system capable of analysis of all ERTS [Landsat] data."[50] This caused problems for the development of a data processing system for Landsat (discussed in chapter 9) because the experimental use of Landsat data required an operational data processing system so that the data would be routinely and promptly available. Landsat supporters came to feel that the Budget Bureau kept limiting funds for Landsat because it expected Landsat to fail to be useful. In the view of NASA managers, the emphasis on Landsat as an experimental project became a Catch-22 situation: The Bureau of the Budget required NASA to build only an experimental satellite but asked the agencies to justify it by its operational usefulness.

The long fight to obtain funding for Landsat caused many difficulties for the project. NASA officials spent much time and energy justifying Landsat and answering criticisms, sometimes the same ones over and over again. Uncertain funding levels made good planning more difficult. As will be seen in later chapters, the problems that plagued the Landsat project were not, except on the most general level, those anticipated by the Bureau of the Budget. Instead, the question of whether the project was experimental or operational became an underlying tension, exacerbated by the Budget Bureau. This controversy had a direct impact on the design of the Landsat technology, particularly the data processing system. However, the blame for the uncertain history of the Landsat project cannot be laid on the doorstep of any one agency.

8
The Design Phase: Organizational and Technological Choices

Once funding for Landsat was approved, the earth resources program office at NASA had to renegotiate its status inside the space agency and its relations with user agencies and other involved organizations. NASA developed both new institutional arrangements for Landsat within the agency and new mechanisms for interagency cooperation. These had to keep pace with the increasing scale of the project and involve the users to the extent that people at the space agency believed to be necessary. Since NASA almost always contracted out design and construction, the approval of Landsat funding also signaled the beginning of major private industry involvement in the project. Each of the firms that examined the possibility of building an earth resources satellite brought its own unique interests and goals to the project. None of these changes involved large-scale disagreements; in fact the relationship between NASA and the user agencies improved. But the many different organizational mechanisms developed in this period were shaped by the same interests that shaped technological choices. The seeds of later problems were planted because many of the new institutional arrangements for the use of the satellite after launch still emphasized demonstration of technological capability, not serving the needs of users.

Management

A number of management problems complicated the normal NASA development process. Upon receiving approval for Landsat funding, NASA managers had to set specifications (make final decisions about what the satellite should be expected to do) and then let a contract to industry to design and build the satellite. However, the Landsat project lacked the strong leadership it needed to undertake these developments and the necessary negotiations with the user agencies. The earth resources program office at NASA Headquarters, which had responsibility for all Landsat policy, suffered a series of leadership and policy changes during the development of the first satellite. While none of its leaders appeared to be bad choices, their short tenure prevented the kind of strong, effective leadership that so often saves innovative projects in bureaucratic settings.

In the fall of 1968 Peter Badgley left his position as manager of the earth resources program and was replaced by J. Robert Porter, a geologist by training who had transferred from the CIA to the NASA policy staff the previous year. Badgley's greatest success had been in dealing with the scientific community; he had the skills needed to encourage the development of enthusiasm for Landsat. However, Badgley had not always had a smooth relationship with higher levels of NASA management. Porter understood better than Badgley the policy problems that needed to be settled before NASA's administrator could support the Landsat program.[1] In particular, Porter possessed the understanding and connections needed to deal with the reconnaissance satellite community and to counter the argument that Landsat should not be funded because it might draw attention to reconnaissance satellites.

Porter handled a key transition, but his tenure at NASA was short; he left in early 1970. Porter decided leave in order to take advantage of an entrepreneurial opportunity that he believed Landsat would provide. He set up a company called Earthsat to provide analysis and special-purpose processing services using Landsat data. Because Porter established Earthsat so early, he and his company played an important part in the development of the role of private industry in the use of Landsat data (see chapter 12).

Porter's replacements, Archibald Park and John M. DeNoyer, were hired from the user agencies, probably in hopes of improving working relationships. Porter's position as head of the earth resources program office went to Park, a veterinarian by training who had made a career at the Department of Agriculture. At the same time, DeNoyer was hired from the Department of the Interior for a newly created post as director of earth observations programs, with responsibilities including weather satellites as well as Landsat. This reorganization helped Landsat; its defenders could link it to the proven benefits of the weather satellite program, and the new office had institutional clout derived from a proven program.

The whole series of changes in management benefited the Landsat project by providing increasing policy sophistication, as the points of view of the intelligence community and the user agencies were brought into the project. However, the rapid turnover of leadership hurt Landsat, because it never had a long-term dynamic leader who would fight for it within the agency and build a loyal constituency. Landsat was not a project shaped by a single dynamic leader, although certain individuals left their mark. Rather, it resulted from bureaucratic development within NASA pushed by the demands of the user agencies.

The Lead Centers

The Landsat project office needed not only to consolidate its position at NASA headquarters but also to create a project team at one of the NASA centers. That project team would develop the detailed specifications for the satellite system and monitor the work of the private company selected to build the satellite. As discussed in chapter 6, NASA top management assigned Landsat to the Goddard Space Flight Center in 1967. Goddard had a group of scientists and engineers who had previously worked on satellites for geophysics but who needed a new project, and NASA leaders perceived Landsat as a scientific project that fit Goddard's emphasis on space science.

The leaders of the project at Goddard brought a coherent vision to the development of the satellite, but their approach also had its own disadvantages. Two individuals played particularly important roles: Landsat study manager Thomas M. Ragland, an engineer who later moved to the Department of the Interior, and project scientist William Nordberg of Goddard's Laboratory for Atmospheric and Biological Sciences. They tended to see Landsat as a scientific experiment first and put practical applications off until later, a view typical of Goddard, which specialized in the development of scientific satellites. Reinforcing this approach was the fact that many of the Landsat personnel at Goddard were transferred from the Orbiting Geophysical Observatory satellite project, which ended after the launch of its last satellite (*OGO 6*) in June 1969.[2] During this project they had gained experience working on a scientific satellite in cooperation with a small group of mostly academic scientists. In contrast, Landsat was a practical satellite being produced for a diverse group of users, many of them technologically naive. The Goddard group lacked experience with developing practical applications of space technology for unsophisticated users.

NASA leaders reorganized the program again in 1972, in preparation for the launch of the first Landsat satellite. At that point the Johnson Space Center received the lead role in the earth resources program, while Goddard continued to serve as the lead center for the Landsat satellite project. The Earth Resources Survey Program Office established in Houston in 1972 served as an "evaluation, system, or staff arm of headquarters." This was part of a larger reorganization intended to reduce the cost of headquarters administration of this and other programs by using the centers to coordinate programs and provide background studies for headquarters decision makers.[3]

NASA headquarters gave the lead center role for earth resources to Johnson rather than Goddard for a variety of reasons. From the point of view of headquarters, giving coordination and policy study to a different center than the one responsible for the main project helped keep one center from gaining too much control over the program. Yet, this division of responsibility did not mean that the newly involved center had to develop the necessary expertise from scratch; Johnson already served as the lead center for experiments involving sensors for studies of the earth carried on Skylab. Giving Johnson the lead role probably reflected the orientation of top NASA administrators toward projects such as Skylab that put people in space and also fit into a larger effort to find new projects to prevent large cutbacks at the Johnson Space Center after the end of Apollo. Finally, personnel at Johnson were better prepared than those at Goddard to work effectively with users. The Johnson Space Center did not have the strong orientation toward basic science that was common at Goddard, and some of the staff of the new program office had extensive experience in dealing with users and transferring technology. Johnson's earth resources aircraft program, in particular, had worked with a wide range of users of earth resources data.[4]

This reorganization had mixed results. It brought earth resources into a higher-prestige position within NASA, but it meant that NASA's earth resources program put more emphasis on experiments carried on piloted spacecraft and gave a lower priority to Landsat. Some users complained that some projects at Johnson to use Landsat data were simply excuses to provide jobs for engineers left over from Apollo, yet the staff at Johnson does appear to have had better skills and contacts for working with users. In the end its success or failure did not matter; NASA discontinued the use of lead centers to supervise and study policy for programs like earth resources in 1977, for reasons that had nothing to do with Landsat. The overall contributions of the NASA centers to Landsat are best judged by case studies, particularly those given in chapters 9 and 12.

Interagency Cooperation

The changing nature of the Landsat project required not only new organization inside NASA but also new organization for interagency cooperation. NASA managers met these needs with a series of new committees, but they kept the role of the user agencies as small as possible by encouraging high-level cooperation rather than cooperation on the working level and by keeping control of the committees they formed.

As the project went into full-scale design, important decisions had to be made that required interagency coordination. Official policy for the project and approval of the specifications given to the contractors were coordinated by a new interagency committee, the Earth Resources Survey Program Review Committee. Formed in June 1968 to improve cooperation between NASA and the user agencies, this group consisted of program leaders such as William Pecora, director of the Geological Survey, and Leonard Jaffe, head of space applications at NASA. The committee met every few months to decide policy for Landsat.

Although some coordination was provided by informal working groups of people from Goddard and the user agencies, NASA policy-makers kept official cooperation at this high level.[5] Goddard scientist William Nordberg complained to headquarters about this system when he learned of it from a memo from Jaffe to Goddard Director John Clark. Nordberg lamented:

We stated the need for establishing a working relationship with user-investigators on ERTS [Landsat] in order to resolve specific problems related to ERTS sensors, instrumentation and data processing techniques. Reference (b) [the Jaffe-Clark memo] instructed us not to establish a formal Working Group as we had recommended, but rather to resolve these problems through informal contacts with the user-investigators. A number of specific problems must now be resolved urgently as some may affect work plans under the Phase B/C study.... Furthermore, certain tradeoffs can be made in the existing sensor design between spectral response, spatial resolution and contrast resolution to achieve a given signal to noise ratio. These tradeoffs depend on and differ with the specific purpose for which the investigators will use the data.[6]

Nordberg's view may or may not have been widespread, but it suggested that at least some of the NASA engineers and scientists working on designing Landsat wanted to give the people who would eventually use the Landsat data a larger role than policymakers at NASA Headquarters allowed.

The Earth Resources Survey Program Review Committee provided an arena for discussions of policy issues by the policymakers of the various agencies, but it gave the user agencies only a limited role. NASA kept control and the right to make unilateral decisions on necessary compromises among the requirements of the user agencies, subject to approval in general form by the Program Review Committee. User agency scientists had little opportunity to participate in making the technical decisions that often shaped policy.[7] Cartographer Alden Colvocoresses at the Department of the Interior's Earth Resources Observation Systems program office in Washington complained:

The existing procedure for interface between users such as Interior and the ERTS [Landsat] program management [Goddard] is not considered adequate. Specifications which vitally effect [sic] the users are being made without the users being properly informed. It is believed that this situation, if allowed to continue, will result in a satellite being flown the data from which cannot properly be utilized by the users.[8]

In addition, the user agencies were represented only by administrators from their research branches. Neither NASA nor the research branches of the user agencies made an effort in the prelaunch period to involve other branches of the user agencies that would be concerned with the operational use of the satellite data.

The user agencies expected a larger role in the project after NASA launched Landsat. In 1972 NASA managers in Washington reorganized the system of interagency cooperation for Landsat once again, to prepare for the new phase of the project after the launch of the first satellite. The Earth Resources Survey Program Review Committee became the Interagency Coordinating Committee: Earth Resources Survey Program, but NASA successfully fought to retain control of the new committee at a time when it might have been more appropriate to give control to the users.

The definition of this new coordinating committee involved not only competition between NASA and the user agencies about control of Landsat planning but also the interests of the Office of Management and Budget in defining NASA as a service agency developing space technology for practical needs (see chapter 7). The budget office wanted the reorganized committee to be chaired by a representative from the Office of Science and Technology, which was the staff of the president's science advisor. The budget office's proposal referred to NASA as the "satellite technology" development agency. NASA managers saw a crucial distinction between that description and their perception of the space agency's role as the "lead development agency for the experimental earth resources program."[9] Administrator James C. Fletcher argued in a letter to Donald B. Rice, assistant director of the Office of Management and Budget, that

> in an experimental program the principal technical agency involved (in this case NASA) cannot be limited to the role of "technology developer on demand" but must have substantive responsibility for overall development and execution of the experimental program. That responsibility, of course, includes the recognition of the user agencies as full partners in the experimental program as well as of their special responsibilities for recommending adoption or rejection of operational systems to meet their own requirements.[10]

The space agency won the skirmish and retained the lead agency role and the chair of the committee. NASA was attempting to maintain its position as the head of the earth resources program not only to keep control of the Landsat project but also to maintain its role as an initiator of technology against criticism from the Office of Management and Budget.

Specifications

Meanwhile, Landsat development proceeded slowly toward the building of a satellite. A wide range of problems repeatedly delayed decisions on Landsat specifications and contractors. These problems stemmed from funding difficulties and the low priority Landsat received both within NASA and in the executive branch more broadly. However, designing and building the technology for Landsat turned out to be considerably simpler than deciding what the satellite should do.

Even before approval of Landsat funding, many potential participants had clear (but varying) ideas about what kind of a satellite they wanted. The Goddard team conducted a concept study of an earth resources satellite in the spring and summer of 1967. RCA, General Electric (GE), Thompson-Ramo-Wooldridge (TRW), and several other companies had already presented unsolicited proposals to NASA for earth resources satellites. The Boeing Company even suggested in 1967 that a refurbished surplus Lunar Orbiter spacecraft could be ready for a mid-1968 launch.[11] Another possibility for an early mission was to fly earth resources sensors on a Nimbus weather satellite or an Applications Technology Satellite. A letter from an engineer at Philco-Ford Corporation to a friend at Goddard suggested one reason why this idea died:

I do not undervalue the impact of the "Udall hoop-la;" that a shining new project is the way to get the most generous funding; and that experimenters and project people prefer the deluxe approach of having their very own satellite. However, I can almost taste the disappointment at seeing the sound technical approach, *and the quickest way to make real progress*, neglected for these political reasons.[12]

The results of the 1967 concept study provided few surprises. NASA officials rejected the idea of incorporating earth resources into an existing project, but they decided that the new project should be based on an existing design for a small satellite like Nimbus. The study also decided what data the sensors should be designed to provide. The resulting plan closely resembled the specifications provided by the Department of the Interior's EROS program in 1966. William Fischer of Interior later complained: "Robinove and I prepared the specifications, data specifications, and NASA then spent a million dollars on the same study that came up with the same answer."[13] As discussed in the previous chapter, NASA's definition of the new project was also slowed by funding difficulties.

Contractors

In May 1969 NASA finally obtained enough funding to ask for proposals from industry for the design of the earth resources satellite the studies had defined. The agency selected two companies to prepare competitive designs in order to encourage economy through competition. This also allowed the space agency to keep some of its technical options open, even at a stage in the process when NASA had to turn over the technical details to a contractor. NASA had decided on compromise specifications and had selected the sensors; industry's role was to plan and build a satellite and a ground system to accomplish the mission. By using two companies in the early stages, NASA obtained two designs and could choose the better one. These concrete proposals considerably narrowed the technological choices available, but the resulting choice still involved a wide range of interests. Within NASA's specifications, the contractors developed their own definitions of the goals of the satellite system, and NASA had to chose both between definitions and between contractors who brought different organizational strengths.

The decisions on specifications already made by NASA simplified the contracts awarded in the fall of 1969. NASA usually divided the design and building of a satellite into four phases: a conceptual systems study (phase A), definition of parameters and trade-off decisions (phase B), systems design (phase C) and development, fabrication, and flight test (phase D).[14] The competitive contracts for Landsat covered phases B and C, since Goddard had already completed phase A. NASA's specifications required the adaptation of an existing spacecraft design, rather than a whole new spacecraft, which limited the companies with reasonable bids to RCA with the Tiros weather satellite, General Electric with the Nimbus weather satellite, TRW with the Orbiting Geophysical Observatory, and Lockheed Missile and Space Company with a modification of an early reconnaissance satellite.[15] After evaluating the proposals, NASA awarded contracts to General Electric and TRW on October 17, 1969.

By the summer of 1970 the work of the contractors had progressed to the point where NASA had to select one of the two competing companies to build the satellite. The most significant difference between the proposals from General Electric and TRW was in the design of the data processing system, which proved to be the most difficult area of technology throughout the Landsat project. The satellite would return 15 million bits of data per second, which would result in a total of 10^{10} bits if it operated continuously during just one of its 103-minute orbits. While NASA never planned that the satellite would operate continuously, over its planned two-year lifetime it would produce an overwhelming quantity of data. This data could either be turned immediately into a photograph or it could be manipulated in digital form by a computer. A photograph allowed corrections to be applied to the image in one operation, for example during the process of printing a video image onto photographic film. On the other hand, digital processing involved the separate calculation of corrections for each point of the image. Digital processing, however, preserved resolution because the points on the image did not get blurred together. The key issue involved in the choice of a contractor was that digital computer processing had the potential to give better results but might be very expensive or even beyond the state of the art for such a large volume of data.[16]

The two potential contractors took very different approaches to this data processing problem. The TRW group chose to use a digital data processing system in its Landsat design, selecting the IBM Corporation as a subcontractor to develop a system to correct the data in a large general-purpose digital computer. General Electric also first considered IBM as a subcontractor but later judged its proposal too ambitious. Instead, General Electric chose Bendix Corporation, which offered a lower-cost plan for a photographic data system. This suited General Electric's corporate philosophy of bidding low rather than offering the most creative new technology.[17]

NASA had to choose between General Electric and TRW on the basis of their complete designs. Both companies' proposals met all the specifications, but TRW's was significantly more expensive, mostly because the data processing system cost more than General Electric's. Some engineers, particularly at Goddard, might have preferred a digital data processing system, knowing that digital processing would be the trend of the future, but others felt that the TRW-IBM system was too far ahead of its time and too complex. It would have been difficult to justify purchasing the more expensive system when both met the specifications and, in fact, TRW's could only marginally produce the required number of photographs.[18] Since engineers expected digital systems to surpass photographic ones in the future, TRW's system might have been more appropriate than General Electric's as the experimental prototype for a future operational system. However, the Bureau of the Budget had required NASA to plan Landsat purely as an experimental system. NASA's selection officials therefore chose General Electric as the prime contractor on July 14, 1970.

TRW protested the selection, for reasons that show both how private industry pursued its own goals and also how the contractual system was inflexible. Goddard had contracted with Bendix for a study of data processing for Landsat. At the time of the Bendix study, General Electric had already selected Bendix as a subcontractor for the data processing system for GE's Landsat proposal. Goddard received the results of the Bendix study, which recommended a primarily photographic data processing system, in January 1970 and transmitted this recommendation to General Electric and TRW in April. TRW charged that Bendix might have provided General Electric with the results of the Bendix study of data processing before those results were available to TRW.

TRW supported its protest by submitting a revised design for the data processing system. The new proposal used three small computers with improved methods of correcting the data, rather than one large computer. In evaluating the new proposal, NASA applications head Leonard Jaffe found that it lacked the relative weaknesses of the original proposal and appeared "to be a very effective and good system."[19] Despite the advantages of TRW's new proposal, NASA officials who reviewed the protest refused to take any action on the company's protest, on the grounds that TRW had not complained about the late report or made any attempt to modify its proposed system until NASA had decided to select General Electric. In addition, while TRW's improved digital system cost no more than General Electric's more heavily photographic system, the TRW satellite still cost slightly more than the General Electric satellite.[20]

Once NASA had selected the contractor, building the satellite proved to be a straightforward process. The only complications arose from the fact that other companies had contracts to build the sensors (see chapter 6), which then had to be integrated with the satellite. Hughes delivered its scanner late, delaying the launch, and the data processing system caused headaches, but no major changes were necessary.[21] The first satellite (shown in figure 7) was successfully launched on July 23, 1972.

Contracts, committees of user agencies, and arrangements of responsibilities within the agency were all part of the process of turning a satellite project from a plan to a reality. NASA managers and engineers were well acquainted with this stage and knew how to make it go smoothly. Two results, however, stand out. The definitions imposed upon Landsat by the Office of Management

Figure 7
Landsat 1. (NASA, 1972)

and Budget put the space agency in a position where it probably had to choose the cheaper data processing technology, even though it had less potential for growth. The agency also failed to establish effective organizations to involve the users routinely in the design process or to involve the operating branches of the user agencies as well as the research scientists. These two sources of future problems stemmed from a failure to recognize that Landsat, because of its practical goals and diverse user community, differed significantly from scientific satellite projects.

9
Data Processing: The Technical Challenge

The most challenging area of technological development for Landsat proved to be the data processing system. That system had to be designed and built, of course, before the launch of the first satellite. In the face of unavoidable uncertainties, the space agency did not choose the boldest technological course for the Landsat data processing system, even in the face of evidence that pushing the state of the art might solve major problems. This conservatism resulted from a compromise between user agency demands, NASA managers' reluctance to take technological risks, and the limitations on Landsat imposed by the Bureau of the Budget. The process of negotiation by which that compromise was achieved provides a good example of how the social construction of technology not only involves choosing between alternative systems but also shapes the design of individual subsystems of a new technology.

Public research and development agencies tend to create more opportunties for themselves by pursuing the most sophisticated technology, but this tendency is counterbalanced by a powerful motivation to avoid risks that might lead to highly publicized failures.[1] In the case of the Landsat data processing system, technological decisions proved difficult to make because neither the requirements for the system nor the available technology was clearly known. Because NASA engineers had to set specifications for the data processing system at the same time as for the satellite, its designers had only a poor idea of who would use the data and in what form the users would want data. In addition, the engineers had to deal with a field in rapid flux, resulting in a constant stream of new technological developments that could affect the system. In such a situation NASA engineers normally took a fairly low risk approach because of the high potential that an innovative approach not only might cost more than planned but also might not work at all. A major failure would result in negative publicity that could affect NASA's appropriations.[2]

Political constraints, particularly those set by the Bureau of the Budget, shaped the choice of a data processing system even more strongly than technological uncertainty. Because of reluctance to commit to an ongoing program, the Budget Bureau insisted that Landsat be a purely experimental program. That is, NASA was supposed to design the Landsat system only to be adequate to test the concept of using a satellite to monitor earth resources, not to be dependable enough for routine practical use. This constraint had a particularly strong effect on the data processing system. An experimental project could afford to have a data processing system that would have to be shut down for maintenance, repairs, and modifications; it did not need to include extra equipment to provide redundancy to ensure smooth operations.

Changing demands then complicated an already limited system. The user agencies provided overly ambitious specifications for what processed data they wanted to receive, and Landsat program managers negotiated with them over what the satellite could provide. These requirements for the system continued to change, however, even after NASA awarded the final contract for the data processing system in 1970. Because funds were not available to change the design as the demands on it changed and because of the limits put on the original design, the data processing system was never adequate to meet the demands. This inadequacy, combined with changes in the pattern of demand for the products and advances in technology, required an entirely different type of data processing system, in which the data were processed entirely by digital computer. NASA managers recognized the need for such a system as early as 1973, and it was finally installed in 1979.

The Nature of the Problem

The goal of data processing is to turn data into useful information. Raw data acquired by the satellite's sensors are transmitted to a processing facility, where they are converted into usable forms, archived in some fashion so that they can be recalled when needed, and disseminated to users. Either these users or some facility shared by several users then extracts information, that part of the data that is useful and was not known before, from the processed data. A color picture of part of Australia is *data*; that picture might reveal the *information* that half the wheat crop in that region had been destroyed by drought. The extracted information in turn has to be archived and distributed.

The Landsat system involved unusual data processing challenges because of the huge amounts of data collected and the wide range of potential uses of that data. The extraction of useful information from processed data can be done by a computer, but this requires large amounts of computer power because each photograph contains so much data and because recognition of significant patterns is often complex. Therefore, in the early years of the Landsat project, information extraction was usually done by a person, a skilled photo-interpreter, looking at the data in the form of a photograph.

The range of potential uses of Landsat data also caused storage problems. It is in theory more efficient to extract the information and discard the unneeded data before archiving and dissemination. This could not be done with most Landsat data, however, because no one could predict what information might later be wanted from the data. In fact, Landsat data had to be stored both as photographs and on magnetic tape because of the varied methods used to interpret it.

Defining the Requirements

Some thought was given to the data processing system during the early discussions of Landsat, but the people involved only slowly realized that Landsat represented a data handling problem unlike any NASA had encountered before. As early as 1967 Alden Colvocoresses, then a staff scientist specializing in cartography at NASA headquarters, suggested that NASA's earth resources program office needed to give more attention to data handling and to urge the user agencies to think about it.[3] By 1986 engineers at NASA's Goddard Space Flight Center had started to plan the data processing system, but the Bureau of the Budget still complained that Landsat data processing was not receiving enough attention.[4] Even after studies started, NASA engineers were slow to recognize how large the demand for the data would be. The problem continued to attract concern even after the awarding of the final contract for the Landsat satellite in July 1970. William A. Anders of the President's National Aeronautics and Space Council suggested in February 1971 that the data reduction task for the satellite as planned might be impossible.[5]

Landsat indeed raised new data processing problems. NASA had handled image data requiring special processing before (for example, from *Lunar Orbiter*). However, that data had in the past come from space probes that collected data for only a very short time. In addition, in most cases the data could be corrected and enhanced for various users at a more leisurely pace after the mission was over. Landsat would return images of as many as 3383 scenes (each 100 by 100 nautical miles) in an 18-day cycle and then repeat the cycle again and again for at least a year. Data therefore had to be processed as quickly as it was received, or the backlog would grow uncontrollably.

The engineers who set specifications for and designed the data processing system for Landsat were caught on the horns of the dilemma of whether they were designing an experimental or an operational system. Most fundamentally, Landsat was an experiment in the systematic use of data about the earth gathered in space, that is, an experiment in information extraction. In order to experiment properly with the systematic use of data, those data had to be routinely and regularly available. In other words, the data gathering and processing system had to be operational rather than experimental. The expense of an operational system, however, could not be justified for a project that might not prove successful. Therefore, the data processing system was built on an experimental scale, not intended to meet operational requirements. This resulted in a system without the redundancy many of the people working on Landsat at Goddard knew the users would need even for experimentation.[6]

In addition to the question of how much capability the data processing system needed to allow thorough experiments in data use, there was the possibility that if the experiments were successful, demand for data for routine use could expand rapidly. This in fact had happened with the first experimental weather satellite, whose data proved to be of such value that they were soon used in routine forecasting.[7] The designers of the Landsat data processing system recognized the possibility that some organizations might make operational use of the experimental data from the satellite, but they considered operational use at most only a secondary objective of the project.[8] Funding constraints prevented planning a data processing system adequate for that eventuality.

Yet scientists and managers at NASA and the user agencies realized that the project would fail without an adequate data processing system. An Earth Resources Survey Program Review Committee document argued for a substantial system within the limitations imposed by the Bureau of the Budget:

A full capability to supply all interested people does not seem to be justified during the early phases of the ERTS [Landsat] experiment, but an adequate capability to supply a wide spectrum of users will be very important for determining benefits that can be obtained from earth resources satellites and for definition and evaluation of follow-on systems.[9]

John DeNoyer, who worked on earth resources satellites at NASA headquarters and later at the Department of the Interior, said in an interview that the system ended up closer to operational than had originally been planned but only because he and others—Archibald Park at the Department of the Agriculture and later at NASA, and William Nordberg at Goddard—told NASA upper management and the Bureau of the Budget that "it just had to be better."[10]

In addition to these fundamental constraints, the design of the data processing system was complicated by the fact that the specifications provided by the user agencies changed as they gained a clearer idea of what the data could provide and for what purposes they could use the data. In early 1969 the scientists at Interior specified that they wanted a quick look at all data and copies of all processed data of land and coastal regions, but they expected to do their own special-purpose processing.[11] The requirements of the user agencies increased as some of the processing facilities they wanted were cut from their budgets and as they became better aware of the potential of the data and the potential difficulty of processing it. The Landsat design team members at Goddard were surprised by these changes because they were used to dealing with scientists working with more limited and easily defined experiments.

In the fall of 1970 the problem came to a head when the user agencies submitted new specifications for the data they wished to receive. NASA had already awarded the final contract for Landsat, making changes difficult. William Nordberg, the project scientist at Goddard, described the Department of Agriculture's request as a "shopping list" and complained: "It seems that not much thought has been given to the relationship between the ultimate objective of the data applications and the capability of the ERTS [Landsat] system. From that point of view alone the Dept. of Agriculture request seems grossly overstated."[12] Interior's request was worse.

Nordberg wrote that "Interior has really sprung a bomb shell on us" by its request for color and precision-processed (extensively geometrically corrected) versions of all scenes taken over the United States, when NASA had planned on providing this special processing for only a small portion of the images.[13] The Goddard group convinced the Earth Resources Observation Systems team at the Department of the Interior that such a large volume of special processing was impossible, and gradually the various agencies reached a compromise on the planned output of the data processing system.[14]

NASA headquarters added to the problem by deciding in June 1970 to add a program of scientific investigators to the Landsat project. As late as early 1970 Goddard estimated that data would be provided to ten users, some of whom would then copy and distribute the data further. The contract awarded that summer did not include plans for a complex data distribution facility.[15] People at NASA gradually realized, however, that an experiment in data use required collection of data in more than a few test zones and expansion of the program beyond the user agencies and the few experimenters who had been involved in the design of the sensors. As described in chapter 10, NASA responded to this need by organizing a program to fund scientists who wished to experiment with Landsat data. This program received over 500 proposals, from which more that 300 experimenters were selected to receive data directly from NASA. These principal investigators would provide more rapid and specific feedback about the quality of the data and the uses to which it could be put than the user agencies would. However, NASA would have to supply the investigators with data because the user agencies could not develop the necessary data distribution facilities quickly enough.

The Goddard Space Flight Center distributed data to the NASA-sponsored scientific investigators and to the user agencies, some of which in turn set up distribution systems to serve other users. The Departments of Agriculture and the Interior and the National Oceanic and Atmospheric Administration built their own data distribution centers to mass produce the standard Landsat data for their own groups of users, using photographic masters provided by Goddard. These data distribution centers and a smaller program at the Department of Agriculture also produced specially processed data for the three communities of scientists and resources planners who dealt with the concerns of the agencies: geology and geography, agriculture and forestry, and oceanography and meteorology.[16] In addition, they served as centers for research into techniques for analysis of the data for the various purposes of each agency.

As long as photographs served as the primary medium of data distribution, it was not too expensive to maintain four separate distribution centers. Landsat managers at NASA hoped that the user agencies knew best how to encourage widespread use in each field. In addition, the Department of the Interior's Earth Resources Observation Systems Data Center (discussed in chapter 11) relieved NASA of a major burden by taking over all distribution to the general public of images from Landsat, Skylab, and NASA's earth resources aircraft program.[17] With the introduction of all-digital processing, however, the expense of maintaining three centers became too high, and the facilities run by the Department of Agriculture and the National Oceanic and Atmospheric Administration were closed in 1977 after complaints from the General Accounting Office about duplication of effort.[18] After 1977 the data center at the Department of the Interior took on most Landsat data distribution responsibilities (see chapter 11).

Technology

The space agency and the contractors faced a major challenge in designing a data processing system to meet these needs. They took an approach stressing practical, low-cost technologies, to an extent that later turned out to be impractical. The system they designed consisted of a primary system that processed all the data, and special processing systems that produced small numbers of computer tapes of Landsat data, color images, and images with more accurate geometric corrections.

A team in the Image Processing Division at the Goddard Space Flight Center planned out a primary, or bulk, data processing system, called the Initial Image Generating Subsystem. In its final form, the steps involved were as follows. As it orbited the earth, the satellite radioed data to ground stations at Goddard, at Goldstone, California, and at Fairbanks, Alaska. The data were recorded by special high-speed tape recorders, and the Goldstone and Fairbanks stations shipped their tapes by mail to Goddard in Maryland. Information on the position and orientation of the satellite was recorded separately. At Goddard, a control computer used the position and orientation information to calculate what corrections needed to be made for the motion and height of the satellite and the distortions of the sensor and to generate the annotations for the finished picture. The next step was to apply these corrections to the tape-recorded data from the sensors, which was done during the process of producing photographs. The control computer directed the electron beam recorder that transformed the tape-recorded data into a photographic image by scanning photographic film with an electron beam to record the image. During this process the computer modified the motion of the electron beam to add the necessary corrections. The resulting photographic master copy was then used to make additional copies and enlargements.[19]

Other sections of the Goddard facility produced color, precision-processed, and computer-tape products. The system produced color pictures by combining three black-and-white negatives of the same scene taken in three spectral bands, which required lining up three different images exactly. This could be achieved fairly easily for the multispectral scanner, which used one mirror to gather information in all the different spectral bands. It was more difficult for the return beam vidicon because it used different tubes for each spectral band, so each of the images that had to be combined had different distortions (see chapter 6).

Precision processing for geometric accuracy and to add map coordinates caused the greatest technical problems. The data received from the satellite had varying distortion within each image, like a photograph printed on a piece of dough and stretched and pushed out of shape, so it could not be corrected by applying a simple formula. Because of the large number of data points in each image, it appeared to be too expensive to have a computer apply the corrections by performing a complex calculation to relocate each individual data point. Instead, the data were manipulated in the form of an image: A special processing machine scanned an enlarged photograph, manipulated the resulting video signal to apply the corrections calculated by the computer, and projected it onto unexposed film. This special processing system was not very satisfactory. The product had accurate map coordinates and better geometric accuracy, but so much of the sharpness and the accuracy of the brightness data were lost that few users wanted it.

Production of computer tapes for Landsat data caused problems not of technique but of quantity. The data received from the satellite was recorded in a very compact form. For most users it had to be calibrated, annotated, and recorded in the less compact (800 bits per inch) standard computer format. In this form one Landsat scene filled four standard computer tapes. The specifications for the data processing facility called for 5 percent of the pictures received from the multispectral scanner to be converted into standard computer tapes.[20] A 1970 memo predicted that this would be adequate because the demand for data tapes would remain small. In fact, rapid improvements in computer technology (discussed later in this chapter) made it possible to process Landsat data on less expensive computers, resulting in a demand for data in the form of computer tapes that quickly outstripped production.[21]

Results

When the first Landsat satellite was finally launched in 1972, the data processing system proved to have a number of inadequacies. The causes of these included technologies that proved less successful than expected, lack of funds, and changing user needs that put excessive demands on a system that was not designed to be very flexible. Landsat supporters had argued that color images and repetitive data would make up for Landsat's lack of resolution, but compromises in design had led to a system that produced only a small number of color images and to delays in processing that made repetitive data less useful.

The users quickly complained about the slow speed of data processing. Although much of the data turned out to be most useful if promptly available, it often took Goddard twenty to forty days to get the data to the principal investigators and to the EROS Data Center for public distribution.[22] Resource managers monitoring snow runoff, range-land conditions, or ice in shipping lanes (to give a few examples) had to respond quickly to various situations. Information on the situation would be useless unless it was received within a few days. NASA had not made any commitment to rapid distribution, and limited funding would have made it difficult to add that to the system. It seemed absurd to the users that data were delivered from Alaska to Goddard by regular U.S. mail, but Landsat produced so much data that any other alternative available at the time would have been very expensive.

The data processing system was not only slow, it was also unreliable. For example, the electron beam recorder that recorded video signals onto photographic film broke down so frequently that at one point it required three full-time engineers to keep it running.[23] John Y. Sos of the Goddard Image Processing Branch wrote in early 1973: "Throughput calculations showed that almost ideal conditions (i.e., high equipment up-time and production efficiency) would be required for processing of the specified ERTS [Landsat] workload. Post-launch operating experience has shown that the conditions were not ideal."[24] Because the users already found the system painfully slow, they strongly resented additional delays caused by equipment problems.

Problems arose not only with equipment breakdowns but also with the effects of standardization on the usefulness of some of the photographic products. NASA had hoped to lower costs and allow easier comparison of images by assembly-line-style production of standardized products, but the standard product could not be ideal for all the various users. For example, so many complaints were received about overexposed images of bright scenes, such as deserts in the summer, that Goddard had to introduce a special product. The original plan to print all the images with a constant relationship between the absolute amount of light received by the sensor and the density of the photographic product simply produced too many images so badly exposed that many users found them worthless.[25] Fundamentally, these problems arose from the fact that the data processing system was designed to produce experimental products for various tests, not the full range of processed data most valuable for operational use.

The growing demand for Landsat data in the form of standard computer tapes meant that the data processing system was soon not only too small and not flexible enough but also based on the wrong approach. The digital data processing system that NASA had decided was too risky in 1970 (see chapter 8) quickly became desirable.[26] As discussed in chapter 10, in the early 1970s many users became interested in analyzing Landsat data by computer instead of by human interpretation of images in photographic form. These users needed data in the form of computer tapes. Large-scale production of such data required advances in the state of the art to make it possible to process the data in a digital computer, which would apply corrections to the raw data and produce computer tapes. The data processing system General

Electric and Bendix had designed lacked flexibility in this area, so when technology for digital analysis improved and the demand for computer tapes increased, NASA had to develop new technology for digital processing of Landsat data to meet the new demand.

NASA managers recognized and began working on this problem quickly, but solutions came more slowly. Soon after awarding the main Landsat contract, NASA requested proposals for the further study of digital processing. IBM and TRW received contracts to develop this technology.[27] IBM produced the first digitally precision-processed Landsat image in October 1973, using a general-purpose computer. As people had feared when digital processing was first proposed for Landsat, the process was very slow.[28] Digital processing and enhancement had been developed by the Jet Propulsion Laboratory (a NASA contract laboratory) to process small numbers of scientific photographs of the moon and planets. Substantial innovation was still required to turn this into a system that could be used economically for the large volume of Landsat data.[29] IBM and TRW continued to work under contracts from NASA to improve the process.

The advance of digital processing technology was fast enough for the Interagency Coordinating Committee for earth resources satellites to recommended an all-digital processing system in its 1973 federal plan and for Goddard engineers to begin to plan such a system in the fall of 1973.[30] The first step was to install a new subsystem to increase the capacity of the data processing facility to produce computer tapes. This subsystem could later become part of a whole new system that processed all the data digitally in a computer instead of handling it primarily in the form of video and photographic images (analog form). The first part of this new subsystem, which transformed the raw data from the satellite into partially processed computer tapes, was installed in the spring of 1975.[31] The rest of the digital data processing system was planned for installation in 1978 at Goddard and the EROS Data Center, but it was delayed until 1980 by lack of funding and technical difficulties.

The early development of the data processing system for Landsat shows the variety of factors that influenced the technology. Changes in the state of the art and in the conception of the project caused the data processing system to be obsolete before it was built. Even though this was recognized, it could not be changed during construction because no money was available for substantial changes in a design that had already been agreed upon and for which a contract was signed. Budgetary limitations and the state of the art shaped the design of the data processing system. So did the previous experience of the engineers and scientists involved, the Bureau of the Budget requirement that Landsat follow an experimental rather than an operational approach, the changing concept of who would use the data, and the sophistication of these users. The technology of the data processing system in turn affected the way the processed data was used, because each type of processed data was most useful to particular users and for certain types of investigations.

III
Application

10
The Transition to Developing Practical Applications

Bringing a new technology into practical use involves both the development of an operational technological system and also the development and diffusion of techniques for routine use of the technology. In the case of Landsat, operational uses for Landsat data had to emerge before an operational satellite system was built in order to justify the need for such a system. The remaining chapters deal with issues that arose during this early phase of the process of developing practical uses for Landsat data. The challenges of this stage of technological change include developing techniques for practical use, integrating them into existing systems, and persuading potential users to adopt the new methods.

Users and the Process of Technological Change

The life cycle of a technology is sometimes conceived as a linear process, from invention or research to development to production and diffusion. In fact, the evolution of most technologies is neither simple nor linear. The diffusion of a technology is crucially shaped by decisions made during research and development.[1] In the later part of the process, the connection is equally close. Many technologies continue to develop even after they reach production and diffusion. In addition, a single technology may have a variety of uses, and those different uses and the necessary subsidiary technologies and techniques may be in different stages of their life cycles. Last, and perhaps most important, few technologies stand in isolation. Most innovations require new institutional arrangements before they can be successfully used.[2]

Thus, to understand the social construction of a new technological system we must understand both how interest groups, including potential users, negotiate the design of that system and also the process of negotiation that determines how that system is actually used. Once a satellite is built and launched, its design has perforce reached some closure—it cannot easily be renegotiated. But the process of defining the application of the data from that satellite is then in its early stages. Technological systems are often used in ways not anticipated by their designers, both because users may find ways of using a new system not predicted during development and because users may resist the intentions of the designers.[3] The use of Landsat data to map shallow water was an example of the first phenomenon; hydrologists examining the data from the satellite found unanticipated information. Users of Landsat data for agricultural survey followed the second pattern, rejecting the techniques NASA had developed for that purpose and developing their own (see chapter 12). More commonly, the development of applications involves some compromise between how the development agency expects the technology to be used and the perceived needs and interests of the users. Thus, uses for a new technological system are negotiated in much the same way as was design of that system, although the interest groups involved and the techniques they use to promote their own interests are not always the same.

In the case of Landsat the negotiation of applications for the new technological system proved particularly troublesome for NASA because of the space agency's mission as a research and development agency, without clear institutional ties to the potential users for the technology it developed. Because the phases of technological change cannot be clearly separated, the transition from research to operational use of Landsat data involved primarily an intensification of existing issues, not a radically new set of questions. NASA managers had had trouble balancing NASA's interests with those of the users even before the first Landsat satellite was launched. These issues continued after launch; for example, NASA and the other agencies involved in Landsat continued to disagree over the proper balance between experiment and routine use of the new system.

However, the interests of the space agency shifted in one important respect after the launch of the first Landsat satellite. The political future of earth resources satellites depended on the development of an enthusiastic group of users who would act as a constituency for the program. Therefore, members of NASA's Landsat team worked to encourage widespread use of the data from the satellite, but they did not always know the best ways to encourage its employment. Promoters of Landsat had to strike a delicate balance between encouraging the operational use of Landsat data and overselling the project before it was ready to fulfill its promise. They had to prove that Landsat was worthwhile while avoiding the danger of prematurely freezing the developing technology by accepting too many operational commitments before the system had fully evolved. Efforts to disseminate the use of Landsat data and the reasons that the results of those efforts were different from those expected by NASA managers are the focus of the remaining chapters.

Persuading users to adopt a new technology is normally an exacting task. In the case of Landsat, diffusion proved particularly difficult because the new satellite technology required significantly more technological sophistication from the users than the systems it replaced, such as aerial survey. In addition, the complexity of earth resources satellite technology meant that research and development and practical use necessarily overlapped and interacted. For the Landsat program the problems of application turned out to be at least as difficult as the problems of research and development. The rest of this chapter examines the scientific research that was the first step in the development of uses for Landsat data. The remaining chapters in this section look at the issues faced in integrating the new technology into various existing systems and encouraging its use. The goal here is not to provide comprehensive coverage of the benefits provided by Landsat but rather to compare the process by which Landsat data came into use by scientists, federal, state, and local governments, private industry, and foreign countries.

Exploratory Research

Scientific research to develop techniques and technologies for extracting information from Landsat data was the essential first step in developing many applications (see figure 8). The availability of the data was not enough; it was necessary to develop techniques for using it to answer particular questions. For example, it was necessary to understand better the geological significance of linear features (like faults) visible in pictures taken from orbit before Landsat data could effectively be used to find oil. A number of difficult issues arose in this process. NASA managers had to decide what kind of research to emphasize: research to develop the

Figure 8
Landsat project scientists at Goddard view an enlargement of one of the early Landsat images. (NASA, 1972)

most sophisticated applications or research aimed at quick practical use. In addition, scientific investigations had to be balanced with dissemination efforts. Early exploration of how Landsat data could be used in different disciplines laid an important foundation for the use of Landsat and set certain patterns, particularly a pattern of emphasizing technological sophistication. In addition, although Landsat program managers realized that scientific research was necessary, they sometimes did not realize that it was not always sufficient to ensure widespread practical use of Landsat data.

NASA officials realized that the program of NASA-supported scientific research would have to be expand with the launch of the first Landsat satellite. Even before the first launch, NASA's earth resources research programs had not only grappled with the technical issue of which sensors provided the most useful data and the scientific issue of how the data were to be subsequently interpreted but also tried to "beat the users out of the bushes"[4] (see chapter 4). However, this effort was limited; scientists and managers at NASA's earth resources program tended to work primarily with research scientists from universities and government agencies. They made little attempt to involve the people who would eventually use the data routinely. However, NASA managers recognized the need to expand this program after a satellite was launched. To prove the value of the satellite, NASA had to encourage development of new methods for extracting information and promote greater interest in remote sensing.

Following the practice established for scientific satellite projects, NASA managers had originally planned to distribute Landsat data exclusively to selected teams of scientists, primarily in the user agencies, and to collect data only when the satellite was over specific test sites. In early 1970 the Landsat project office at the Goddard Space Flight Center estimated that processed data would be provided to only ten users, and contracts awarded that summer made no provision for a complex data distribution facility (see chapter 9).[5] However, with a push from the EROS personnel at the Department of the Interior, NASA officials came to realize that experimenting with the use of new kinds of data required the collection of data in more than a few test zones. Landsat data had the potential for many different kinds of uses, and as many as possible of these had to be developed in order to justify the project. The program would have to be expanded beyond the experimenters at NASA and the user agencies who had been involved in designing the sensors.

NASA Headquarters decided in June 1970 to request proposals for the use of Landsat data from scientists throughout the world, so that the data would be tested for a wide variety of uses. These researchers from outside the group of agencies already involved in Landsat were called Landsat principal investigators. Unlike the principal investigators for NASA's scientific satellites, however, they did not participate in the design of the instruments or have exclusive rights to the data, even for a limited period of time.[6] The Landsat principal investigators received free data and, in most cases, funding in return for reporting to NASA the results of their experiments. NASA hoped that these scientists would prove the value of the Landsat data and develop specific applications.

Once the decision to include principal investigators was made, the number of investigators grew rapidly. In an early discussion of the possibility of including outside scientists, in November 1969, NASA Headquarters unofficially estimated that 40 to 50 principal investigators would join the program. It turned out that scientific interest in Landsat was greater than NASA officials had realized. The request for proposals issued in mid-1970 elicited over 500 proposals. Hoping to encourage the broadest possible use of the data, the Landsat project office attempted to fund, or at least provide free data to, as many of these investigators as possible. NASA set up peer review panels in each of the various disciplines (chiefly geology, hydrology, and geography), which chose over 300 investigators to receive support from the space agency.[7]

The program of principal investigators raised an important issue of data distribution policy. NASA officials decided that it was necessary to release all Landsat data to the public at the same time as it was released to the principal investigators and user agencies. As noted previously, this contradicted the usual agency policy of allowing principal investigators exclusive rights to their data for a few months so that the individuals who had invested their time and effort planning an experiment could publish the first results. Landsat data had to be treated differently, however, because some information could potentially be exploited for personal economic gains. For example, if a Landsat principal investigator discovered evidence of oil deposits on land thought to be worthless and obtained the land at a low price, the previous owner of the property could argue that NASA, a government agency, had given the investigator, a private citizen, an unfair advantage. To avoid this, NASA decided to release all data to the general public immediately. Theoretically, this would give everyone an equal opportunity to put Landsat to work.[8] By adhering to this policy, NASA and the Department of the Interior occasionally frustrated Landsat investigators, who had to wait for their data to be catalogued for public release. However, NASA believed that this policy not only prevented scandals but also encouraged wider use of the data.[9]

Basic Science and Practical Science

Landsat produced valuable scientific results starting from the launch of the first satellite in July 1972, but NASA officials soon realized that the research done by the research arms of the user agencies and the scientists participating as Landsat principal investigators was a long way from practical use. The earth resources program initially tried to move research toward applications by supporting different kinds of research. In some cases this plan succeeded, but even the new program went only partway. It emphasized demonstrations of the practical use of Landsat data, not efforts to actively persuade potential users to adopt the newly demonstrated techniques.

Initial results from Landsat proved promising, although the scientists had many frustrations and disappointments. Used to getting top priority in NASA's scientific programs, scientists complained bitterly when the NASA policy of public release of data and the data processing problems discussed in chapter 9 hindered their work. Despite these problems, reports presented at the 1973 Symposium on Significant Results at the Goddard Space Flight Center showed that data from the first mission quickly provided both scientific results and practical potential as well as a few disappointments. Geologists found many new large-scale features, particularly faults and other linear structures, some of which were known to be useful for finding oil. Some specialists had hoped that the data could be used to classify types of rocks, but that effort failed. Agricultural scientists discovered that crops could be identified, at least in small regions, and the areas planted could be measured. It became clear, however, that predicting countrywide or worldwide harvests would require an extensive study of regional variations and a tremendous amount of work on computer classification and analysis (see chapter 12). Hydrologists found the Landsat images more useful than expected for mapping water areas, both normal and flooded, and for monitoring snow cover and glaciers.[10]

NASA managers assumed that these scientific results from Landsat could be quickly translated into practical applications. In order to encourage practical use of the data, the Landsat project office followed a different strategy in 1974 in selecting the second group of Landsat principal investigators, who were called Landsat follow-on principal investigators. A study of the *Landsat 1* investigations had revealed that few of the investigators had worked with resource managers, the ultimate operational users of the data. Nor had many investigators provided statistical or comparative analyses of the practical value of the techniques they had developed.[11] For the next set of investigations, the Earth Observations Office at NASA Headquarters funded a smaller number of researchers with a more practical orientation. In selecting the fifty-five follow-on investigators, the Landsat office at NASA Headquarters and the Earth Resources Program Office at the Johnson Space Center looked not for scientific breakthroughs but for demonstration projects that developed operational uses for the data.[12] The follow-on principal investigators received their data from the Department of the Interior's Earth Resources Observation Systems Data Center (see chapter 11) or from the much smaller data distribution facilities at the Department of Agriculture and the National Oceanic and Atmospheric Administration.

Operationally oriented research proved to be more difficult to identify and support and also to conduct successfully. Developing practical uses for the data required not only developing techniques for using the data but also integrating those techniques into existing systems. Clifford E. Charlesworth, head of the Earth Resources Program Office at Johnson Space Center, complained that "the quality of the proposals was not as high as we hoped," and he had to fight to preserve adequate funding for the selected follow-on principal investigators.[13] Stanley C. Freden, a scientist in the Missions Utilization Office at Goddard, wrote in 1975:

> I don't feel that we have yet made a successful effort in getting potential users into the program. The emphasis on quasi-operational demonstration projects for the follow-on investigators was a first step but for many of those investigations one has to stretch his imagination to put them into that category. I feel that we can go much further in that direction. There are thousands of potential users out there and we must reach them. Most of them will require some "seed" money to help them get started.[14]

One problem arose from the high level of technology in the research NASA had supported. Research conducted by principal investigators, NASA, and the user agencies had resulted in many sophisticated techniques for applying the data, particularly by means of computer analyses. Specialists wrote computer programs for pattern recognition and classification, so that data could be analyzed automatically without the traditional assistance of large numbers of skilled photo-interpreters. NASA encouraged this trend toward sophisticated digital analysis, which allowed more information to be extracted, albeit at greater expense. But this was not necessarily what the users needed most nor was it the best approach to promoting the use of Landsat data.

NASA administrators knew how to support scientific research but pushing that research toward useful results was more difficult. In the first few years most research was necessarily more scientific than practical. By the mid-1970s, however, NASA believed that the scientists it had supported had developed applications of Landsat data that were ready for operational use in a number of fields. In most cases this proved to be difficult, as illustrated in chapters 12 through 15.

Moving from Science to Application: Hydrology

Before analyzing the major Landsat applications projects, it is valuable to examine some simpler cases that illustrate many of the issues involved. The contributions of Landsat to the study of lakes and shallow ocean areas turned out to be unexpectedly significant and useful. The techniques developed to use Landsat data moved fairly quickly into practical use because the data provided needed information that had not been available before, rather than replacing existing methods of gathering information.

Landsat 1 hydrological investigators monitored areas covered by water and snow and attempted to measure water quality. Unlike scientists in other fields, hydrologists found that Landsat data could be used more easily for their scientific investigations than they had expected. Snow and water stood out plainly on Landsat pictures, although snow had to be differentiated from clouds, and water from cloud shadows. Shallows and patterns of turbulence showed clearly in large bodies of water. While Landsat clearly provided data of interest to hydrologists, extracting information from that data and developing systems for using it still required considerable research. For example, mapping snow and water areas turned out to be only part of the task of using Landsat data to predict spring flooding. Scientists also had to develop new theories and models to relate measurements of the area covered with snow (with no information about the depth of snow) to spring runoff. Estimates were needed of the volume, rather than just the area, of lakes. In other cases, details visible within water areas in Landsat images had to be correlated with depth, suspended sediment, floating algae, and pollutants.[15]

Mapping surface water was a popular use of Landsat data. The Earth Resources Program Office at the Johnson Space Flight Center conducted a joint project with the Army Corps of Engineers and the Texas Water Rights Commission and Water Development Board to develop a computer program that used Landsat data to automatically identify bodies of water larger than about one hectare (a little more than two acres).[16] This work attracted widespread interest after several serious dam accidents prompted Congress to order the Corps of Engineers to inventory all dams in the United States. As a result, NASA transferred its experimental surface-water detection procedure to the Corps in 1975.[17] Unfortunately, the program identified too many shadows as bodies of water for the engineers to make efficient use of the technique on a nationwide scale. For example, a March 1976 analysis of one Landsat scene of Washington state located 333 bodies of water not included in the existing dam inventory. Of these, 324 were natural lakes, standing water, or shadows. The remaining 9 were ponds created by dams, of which 3 were large enough to be included in the dam inventory.[18] Despite the problems with this procedure, research continued on improved techniques for data interpretation. By 1979 the technique was operational; the Corps of Engineers routinely used Landsat data to update the dam inventory.[19]

Hydrologists also developed a method for charting water depth from Landsat data. Cautious at first about using satellite data for bathymetry because water clarity varied widely, scientists found early research to be more promising than expected.[20] With an eye for good publicity, NASA let a study contract to well-known environmentalist Jacques Cousteau in a cooperative project with the Environmental Research Institute of Michigan. Their work demonstrated that Landsat could be used to chart water depth to twenty two meters with an accuracy of 10 percent in clear water.[21] In early 1976 these results attracted the attention of the Defense Mapping

Agency of the Department of Defense, which requested Landsat images of certain critical shallow sea areas. Just a few months after receiving these data, researchers made an important discovery: a new reef in the Chagos Archipelago in the Indian Ocean.[22] Subsequent results proved so useful that in 1978 the Defense Mapping Agency requested Landsat coverage of coastlines and shallow seas for most of the world.[23] Like the Central Intelligence Agency, the Defense Mapping Agency sometimes chose to use Landsat data for certain mapping tasks because it was more economical than data from classified satellites. Unlike classified data returned by military satellites, Landsat data did not require expensive special handling.[24]

Hundreds of examples of successful and unsuccessful tests of uses for Landsat data could be described here to illustrate the point that scientists discovered a large amount of information in the new data. However, it gradually became obvious to NASA managers that scientific investigations did not necessarily lead, without further effort, to practical applications that would be adopted by resource managers. The agency had to balance the needs of the scientific constituency it had created with those of the operational community it hoped to spawn. The two examples just discussed show relatively smooth transitions from scientific to practical use of the data. The next five chapters examine in detail cases in which the transition went less smoothly. In those controversies the process of negotiation among interest groups became particularly clear.

11
The Department of the Interior's EROS Data Center

To examine the social construction of operational use of Landsat data, it is necessary to examine not only NASA initiatives but also the role of the Departments of Agriculture and the Interior in bringing Landsat data into routine use. This study involves two questions: How widely and how rapidly did the user agencies come to employ earth resources satellite data to meet their own missions? Did these agencies also play a role in promoting Landsat to potential users outside their own organizations? This chapter examines the role of the Department of the Interior, which made a significant effort to promote widespread practical use of Landsat data. As in the design of the satellite, the interests of the Department of the Interior were reflected in the types of arrangements for operational use it promoted. Its interests included promoting the data distribution center it had set up and serving an existing constituency. Many of the resulting ventures were effective; Interior may even have been more successful than NASA in encouraging the diffusion of techniques using Landsat data. The following chapter deals with the rather different case of the Department of Agriculture, which was much less interested in Landsat and had to be pushed by NASA to experiment with satellite data.

Initially, Interior leaders wanted their own earth resources satellite program (see chapter 5). When it became clear that NASA would keep control of the experimental satellite program, Interior sought a major role in the area of data distribution and analysis. Despite chronic funding problems, the Department of the Interior developed an integrated program to promote the use of earth resources satellite data and built a data distribution center. The Department of the Interior's Earth Resources Observations Systems (EROS) Data Center eventually took over from NASA most distribution of Landsat data. Because of the shortage of funds, politics played an especially large role in the development of the data center. This made the success of the data center in promoting Landsat data use particularly significant: The negotiation among interests to define the use of a technology cannot be interpreted as a contest between harmful political interests and beneficial interests that promote technological efficiency.

Building a Data Center

The group involved in the Department of the Interior's attempt to gain permission to build its own earth resources satellite continued to exist as the EROS program even after the White House denied authorization for an early Department of the Interior satellite (the change in focus resulted only in a change of the meaning of the acronym from Earth Resources Observation Satellite to Earth Resources Observation System). Under the leadership of William Fischer, this group organized itself around the expectation that Interior would eventually assume responsibility for operational earth resources satellites just as the Weather Bureau had for meteorological satellites. Given this expectation, it seemed logical to Interior personnel that their department should organize the system for Landsat data distribution from the start and then take over other functions when the program became operational. They therefore gave a high priority to getting permission to build a major data distribution center. Their effort succeeded only with the help of pork-barrel politics.

In 1968, while still trying to persuade NASA to make a quick transition from the experimental Landsat project to an operational earth resources system, EROS personnel at the Department of the Interior began to plan a data distribution system. The Geological Survey, of which EROS was a part, let a contract to RCA for a "Ground Data-Handling System Study for the Earth Resources Observation Satellite." The RCA study assumed an earth resources satellite that would carry a return beam vidicon and a ground data processing facility that would use an RCA Laser Beam Image Reproducer to produce Landsat pictures.[1] The study thus assumed that NASA would build the satellite that the Department of the Interior had proposed when the EROS program was announced in 1966 (see chapter 7).

The RCA report provided a plan for a data distribution center; the next step was to select a site and obtain funding. RCA recommended that the Department of the Interior locate the data center where it could receive transmissions directly from the satellite when it was over any part of the continental United States. This limited the possible sites to a small area in the northern Midwest. In addition, later studies specified that the center should be located away from possible sources of radio interference, ruling out sites near large cities and military bases. The Geological Survey issued a contract for the selection of a site in late 1969.[2] However, the Department of the Interior had not yet had much success in gaining funding for the EROS program (see chapter 7), so site selection was going on at the same time as the processing of obtaining authorization and funding for a data distribution center.

The process of choosing a site showed local-development and pork-barrel politics at work. Congressman Ben Reifel and Senator Karl E. Mundt of South Dakota heard of the search for a midwestern site and urged community leaders in various cities to suggest South Dakota as the ideal location. The city of Sioux Falls made a bid, organized by local business leaders David Stenseth, Al Schock, and Russ Pohl. Pohl represented local high-tech industry; he served as a vice president of Raven Industries, which made, among other products, high-altitude research balloons. When Sioux Falls was selected as one of the candidate sites the campaign intensified. The Sioux Falls Industrial Development Foundation produced a two-volume report, "This is Sioux Falls, South Dakota," and entertained visiting scientists, managers, and politicians.[3] Sioux Falls community leaders saw the data center as a particularly desirable catch because it would generate publicity for the city and increase the local level of technology.[4]

As a result of local efforts in Sioux Falls and Senator Mundt's political bargaining power with the Nixon administration, Sioux Falls had an extremely strong position. While the Department of the Interior was deciding on a site, Mundt sponsored an amendment adding $300,000 to the Department of the Interior's budget for the data center. When the Bureau of the Budget withheld the money, Mundt went to President Nixon and secured its release. Cooperation benefited both Mundt and the Department of the Interior: Mundt secured the data center for Sioux Falls and thus brought economic benefits to his constituents, and the Department of the Interior obtained funding for the center that would otherwise have been denied. Meanwhile, the Sioux Falls Industrial Development Foundation offered to donate land for the data distribution facility to the federal government and to construct a building for the Department of the Interior as part of a lease-purchase agreement.

The political nature of the final decision raised controversy and suspicions. After a direct appeal by Mundt to the president for final approval of the decision to build a center in Sioux Falls, the senator announced on March 30, 1970 that the Department of the Interior had selected Sioux Falls as the site for the data center.[5] Senator Carl Curtis of Nebraska learned only after Sioux Falls had been selected that the RCA report had equally recommended Lincoln, Nebraska. He protested to Secretary of the Interior Walter J. Hickel that "the State of Nebraska has not received the fair and proper consideration by the Department of the Interior."[6] While Senator Curtis did not succeed in forcing a reconsideration of the decision, the whole process left many people in the federal government suspicious of the political character of Interior's handling of the earth resources satellite program.

The center received strong local support, but the local-development process also reflected competing interests. Within three months after the site selection, the Sioux Falls Development Foundation had raised more than the necessary $390,000 for the land and building.[7] However, at a hearing in August 1971 about protective zoning around the data center site to prevent radio interference, some farmers complained vociferously about restrictions on the use of their land. The minutes of the next meeting of the directors of the Industrial Development Foundation reported that:

Pohl, not being a native South Dakotan, expressed his belief that he had always told people and had always believed that the people of South Dakota were down to earth GOOD people. A meeting such as the one held Monday evening, said Pohl, he would never have believed could happen in South Dakota. R. Pohl said that he thought the upset was caused by 1) Greed and 2) Ignorance. The few people causing the disturbance wanted a guarantee that their land value would go up.[8]

The controversy was quickly smoothed over, but it showed the costs of relying on local political support.

The political support of Senator Mundt did not save the EROS Data Center from further budget problems. Apparently, the Nixon administration was more willing to make promises for political reasons than to follow through with funds. Data center managers became experts at creative financing, for example, using a subsidy from the Work Incentive Program to hire production workers off the welfare rolls.[9] Funding problems delayed the groundbreaking for the new center until April 1972 (see chapter 7). This put the Department of the Interior in a difficult position, since the data distribution center would not be ready in time for the launch of the first satellite.[10] The EROS program, determined to take a central data distribution role, opened a data center in temporary quarters until the building was completed. The temporary facility was inadequate: Freezing pipes caused a serious flood, and overcrowding was so severe that a photographic processor was put in the only women's restroom.[11] The permanent facility, the Karl E. Mundt Federal Building, opened in August 1973, more than a year after the launch of *Landsat 1*.

Operation of the Data Center

In order to set up a successful distribution system, the EROS Data Center had to surmount both organizational and technical problems. It inherited a tense relationship between NASA and the Department of the Interior. Difficulties in communication between Interior and NASA added to technical problems that could be traced to tight budgets and high demands. Some of the uncertainties were inevitable in a new project, but others grew from competing interests and from the process of trying to build an operational data distribution center for a project still firmly limited to an experimental mission. For example, NASA managers doubted (with, at first, some justification) the technical expertise of the engineers at the EROS Data Center.[12] Indeed, the Data Center sometimes lacked enthusiasm for, or could not afford, the most sophisticated technology for data analysis. However, personnel at the EROS facility often were more effective than NASA staff in trying to think about the needs of the users.

NASA's interest was not to control all Landsat data distribution; NASA managers willingly shared certain data distribution tasks with other organizations. Most notably, NASA did not want to sell Landsat data to commercial customers because the space agency would not get to retain the money received but rather would have to turn it over to the government's general fund. The Department of the Interior, on the other hand, had authorization to retain the proceeds from the sale of data and information products like maps to offset the cost of producing them. NASA managers were therefore delighted to assign to the EROS Data Center the job of distributing data to the public.[13] Robert L. Bell, a policy expert at NASA Headquarters assigned to technology utilization, wrote:

> The policy I advocate is that if we feel that someone can contribute to the program in any way we should supply the data to them without cost. The business that I am not anxious for us to undertake is supplying copies of the imagery to the curious, to the high school students, and to the petroleum exploration people, who most assuredly are not going to tell us what they have done with the imagery. This is the job I want to give to Sioux Falls.[14]

The space agency also assigned to the EROS Data Center distribution to the public of data collected by NASA's high-altitude aircraft. Initially, however, NASA was not willing to turn over all Landsat data distribution to the Department of the Interior's Data Center. The space agency retained responsibility for data distribution to the scientists it funded and supported separate data distribution facilities at the Department of Agriculture and the National Oceanic and Atmospheric Administration.[15]

The EROS Data Center was also shaped by lack of funding to build the kind of facility that Department of the Interior leaders had planned. Data Center managers had to work out a division of data processing responsibilities with NASA's Goddard Space Flight Center, which received the raw data from the satellite and distributed data to NASA's principal investigators until 1975 (see chapter 9). The EROS group had hoped to take responsibility for putting the data into useful form, as well as for reproduction and distribution. However, the Department of the Interior could not get authorization until 1977 for a facility that was criticized as a duplication of data processing capability already available (if marginal in capacity) at Goddard. Since the Data Center lacked the facilities to process the data at the center, the EROS people wanted some control over the type of processing done by NASA of the data that the EROS Data Center would distribute. NASA, in turn, thought the Department of the Interior's initial requests for processing unreasonable (see chapter 8).

Within the limits of available funding, Data Center staff set out to design the system to serve the needs of an eventual operational satellite system. An operational facility had to be set up for mass production; that is, it had be designed not only to be able to handle large-scale production but also to ensure the smooth flow of products through the system (see figure 9). Meanwhile, the designers of the system did not yet have any practical experience on which to base predictions of how much of what kinds of data users would want. This made designing an indexing system to select the best aircraft and spacecraft imagery to fill a user's request a major challenge.[16] Quality control was also a recurring problem, particularly because different users had different definitions of top-quality imagery. For example, some users wanted a picture on which features could be seen most clearly, whereas others wanted all images identically printed so that actual changes were not masked by photographic enhancements designed to produce a better-looking picture.[17] Because it was a mass production facility rather than a custom laboratory, the Data Center had to compromise. In fact, when the Data Center did offer special processing, it often received complaints about government competition with private industry from companies that provided custom processing services (see chapter 14).

Figure 9
Production of Landsat data in the form of photographic prints at the EROS Data Center. (Department of the Interior)

Eventually the production process reached smooth operation, but the Data Center managers were disappointed by the customer response. Total annual production stayed under 400,000 frames, much less than the 1 to 2 million frames per year predicted in prelaunch studies. Most important, the demand for different types of data did not match prelaunch estimates; predictions of the needs of users had not taken into account the shift toward computer analysis of data. As indicated in figure 10, sales of Landsat images by the EROS Data Center held steady, rather than growing rapidly as had been predicted, while figure 11 shows significant growth in the sales of data in the form of computer tapes. Figure 12 shows the distribution to the several classes of data users. Use of Landsat by state governments and private industry is discussed in chapters 13 and 14.

Disappointing sales of Landsat data resulted from at least four factors. First, many of the early customers were simply curious, a market that declined after the first spate of publicity waned. Second, most operational users of Landsat data in the 1970s used Landsat images as maps or for geological studies and therefore needed to acquire only one set of data. Not many early users used

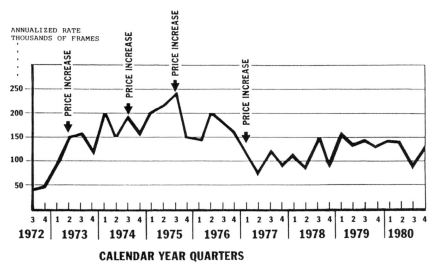

Figure 10
EROS Data Center production of images.

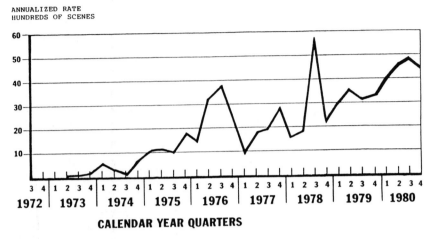

Figure 11
EROS Data Center production of computer tapes.

Landsat data to look at seasonal and continuing resource changes. Third, when the computer hardware needed to analyze Landsat data became more routinely available, many major users switched to buying computer tapes, which they selected with more care because tapes were much more expensive than photographs. Fourth, some early users found Landsat data disappointing and felt that in their enthusiasm the project's supporters had promised too much.

Encouraging New Users

Disappointing growth in sales of Landsat data encouraged the EROS Data Center to plan efforts to encourage operational use of Landsat data. Data Center managers did not tackle the whole problem of finding more customers for Landsat data, but they did encourage and assist certain types of users. The Data Center's mission clearly included research and training for various branches of the Department of the Interior, which also relied on the Center to supply them with Landsat and aircraft data. The EROS Data Center therefore concentrated its training and sales efforts on two classes of users: the various branches of the Department of the Interior and foreign users.

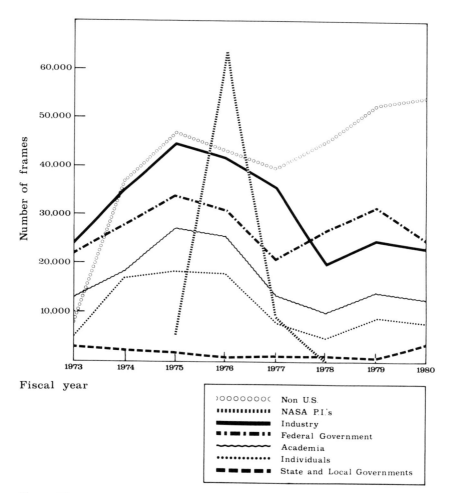

Figure 12
Types of EROS Data Center customers for Landsat data in photographic form.

EROS personnel had very limited resources for research and for training prospective users, but these activities were recognized as important from the beginning. The EROS Program Office in Washington (later moved to Reston, Virginia) conducted research in cartography (under the leadership of Alden P. Colvocoresses), in standardizing categories for land-use mapping, and in other fields. Initially, the EROS Program Office viewed research as its role, whereas the Data Center would provide training to users.[18] With time, however, the Data Center increased its role in research on applications and data processing and limited its training function.

The Data Center's efforts to train users and promote the Landsat program differed in a number of important respects from NASA's program. EROS Data Center scientists gave particular attention to developing ways of integrating Landsat data into large-scale information systems. They sought to encourage the development of collections of a wide variety of data that a user could query by computer to obtain the particular information needed for a given local study. For example, resources managers could monitor forest conditions more effectively if weather information and soil maps were available along with the Landsat images. In comparison, NASA scientists tended to emphasize the value of Landsat and to show less enthusiasm for using Landsat data as a small part of a large information system. Data Center director Allen Watkins criticized the NASA program as "highly promotional and heavily subsidized" and contrasted its emphasis on using primarily Landsat data with the Data Center's programs, "which are discipline-problem oriented and stress seeking solutions to problems using any sources of remotely sensed and ground based data."[19] Users tended to perceive the EROS program as more sympathetic to and understanding of practical and political needs than NASA.

As a part of the Department of the Interior, the EROS Data Center staff naturally gave first priority to training and technology transfer within the department. The Data Center drew its civil service staff from the Geological Survey, but the EROS program officially provided remote sensing expertise not only to the Geological Survey but to all branches of the Department of the Interior.[20] Providing expertise, however, turned out not to be enough to bring Landsat data and the services provided by the EROS Data Center into widespread use within the Department of the Interior. Although the scientists in the user agencies who worked with NASA on Landsat, such as the personnel of the EROS program, were enthusiastic about Landsat from the start, the resource managers in the operational bureaus of those same agencies were reluctant to use the new data. One user agency scientist explained that "the people with operational program functions already have means to accomplish their ends and have to be convinced of the advantages of ERTS [Landsat] data."[21]

The Data Center's promotion and training efforts successfully encouraged use of Landsat data by some of Interior's departments. By 1976 the Bureau of Reclamation was planning to acquire its own computer analysis system to work with Landsat data, primarily for land-use classification. The Bureau of Land Management had similar plans for an automated information system. In mid-1976 the EROS program manager wrote that the Bureaus of Reclamation and Land Management "appear to be at the 'take-off' period in their adoption of remote sensing, particularly digital image processing."[22] The description that the bureaus were in a take-off period suggests that EROS program managers were not only promoting Landsat data use but also explicitly analyzing alternative strategies for that purpose.

The Data Center also had some success in working with other bureaus of the Department of the Interior that were not as advanced in Landsat data use. In 1976–77, for example, the Data Center conducted basic training courses for employees of the Fish and Wildlife Service. The Data Center promoted these programs by emphasizing user needs. For example, when EROS Data Center Chief Allen Watkins discussed with the Fish and Wildlife Service the possibility of conducting a more advanced course on digital analysis, he wrote, "Further discussion of the needs of the Fish and Wildlife Service will be necessary to tailor that course to your requirements."[23] This approach was reasonably successful: A questionnaire determined that the use of remote sensing data by Fish and Wildlife and Land Management personnel who took a course at the Data Center rose from 20 percent before training to 75 percent one year later.[24]

These programs trained a substantial number of Department of the Interior and foreign scientists as well as some representatives of private industry and personnel from other government agencies, who participated in courses at the Data Center when space was available. In fiscal year 1976 the EROS Data Center conducted workshops and training courses for 178 Department of the Interior employees, 63 employees from other federal agencies, 170 state and local government employees, 130 foreign participants, and 179 trainees from such organizations as the InterAmerican Development Bank, the United Nations, and the American Institute of Mining Engineers.[25] NASA took over training of state and local personnel in 1978, which left the Data Center with federal and foreign students. This still involved a larger demand for training than the EROS facility could meet. In fiscal year 1978 the Data Center conducted twelve training courses for Department of the Interior bureaus, five courses for foreign scientists, and one workshop with participants from the Department of the Interior, state agencies, and private industry on using remote sensing to monitor strip mining.

The courses for international students grew from the Geological Survey's tradition of cooperative work with foreign countries. Such work provided both a public service and an opportunity for research that could not practically be conducted in the United States. For example, in the United States, private companies would complain of government competition or favoritism if EROS scientists developed techniques for oil and mineral detection, since private enterprise normally conducted such research. In foreign countries that had government-owned extractive industries, however, problems of competition with private industry did not exist. In addition to research, the EROS Data Center conducted four-week international courses twice a year and occasionally provided courses outside the United States, at the host country's expense. By early 1981 EROS Data Center scientists had conducted 31 training courses for 859 non-U.S. students from about 80 countries.[26] Demand remained high; in 1978, for example, there were two or three times more applicants than spaces available.[27]

The staff at the EROS Data Center followed a different training strategy than did NASA programs. Particularly in the courses for foreign students, the Data Center training program stressed simple manual analysis of Landsat and aircraft images. The Data Center avoided advocating the use of advanced technology, particularly computer analysis—which required large (on the order of $0.5 million) investments—except for those users for whom it was essential. William Draeger of the Data Center training program explained that government agencies and bureaus came to the center with an information need that they wanted filled as simply as possible; it was not the role of the Center to tell the agencies how to do their job or what information they needed.[28] A sizable proportion of the data stored at the EROS Data Center consisted of Department of the Interior and NASA photographs taken from aircraft, and studies sometimes concluded that the best data source for a particular user was aircraft photographs alone rather than Landsat images. The head of the Applications Branch at the center, Donald Lauer, noted that foreign students in particular tended to be enamored of digital analysis just because it was the most sophisticated technology. To counteract this tendency to use more advanced technology than was necessary, most of the courses did not introduce digital analysis until the students had mastered photo-interpretation. The Data Center staff sought to encourage practical uses of Landsat data instead of providing training in sophisticated but impractical techniques.

The EROS Data Center had a history of pork-barrel and self-serving interagency politics, both in its founding and in the role it played relative to NASA. Despite this, the center took a valuable intermediate role between the technology developer, NASA, and the ultimate users. By choosing a manageable problem, the Data Center had more early success than NASA in encouraging the use of Landsat data. The other function of the center, data distribution, suffered from the problems of trying to build an operational capability for an experimental program and from lower than expected demand. Despite these problems, the data distribution facility was successful enough that NASA eventually turned over almost all data distribution functions to the Department of the Interior. Gradually, the Center assumed data processing responsibility as well, but meanwhile it had made a place for itself in user training.

12
Agricultural Applications of Landsat Data

Early cost-benefit studies of earth resources satellites suggested that the largest benefits would come from using satellite data for agricultural inventory and detection of diseased or damaged crops. NASA set out in 1972 to develop these capabilities in a major project called the Large Area Crop Inventory Experiment (LACIE). The experiment succeeded in predicting the yields of certain crops in certain parts of the world. However, the Department of Agriculture did not adopt the LACIE system on an operational basis, as NASA leaders had planned. The system did not match the interests of the Department of Agriculture, and NASA managers failed to cultivate successfully potential users at the Department. However, one branch of the Department of Agriculture did adopt routine use of Landsat data, developing its own system based on a very different approach from the one advocated by the space agency. This case provides an interesting study of the ability of organized users to redefine a technology, despite the best efforts of the developing agency to encourage a particular kind of use.

LACIE was an experiment in using Landsat data to predict crop yields. NASA organized LACIE as a project involving the Department of Agriculture and the National Oceanic and Atmospheric Administration. The goal of the project was to predict wheat harvests in various countries, such as the Soviet Union, by using Landsat data to measure the total area planted in wheat and then making predictions based on meteorological data to estimate the yield from that area. NASA argued that this information could produce financial benefits in two ways: savings to the Department of Agriculture on the cost of agricultural surveys in the United States, and higher profits on wheat sales to foreign countries by American farmers and grain dealers who had advance knowledge of poor harvests elsewhere.

Origins of LACIE

NASA's first large-scale agricultural experiment with remote sensing was a research project called the Corn Blight Watch. During the 1970 and 1971 growing seasons, NASA's Johnson Space Center, in cooperation with the Department of Agriculture and Purdue University, used high-altitude aircraft to detect corn blight, which was unusually severe in those years (see figure 13). Blighted or otherwise damaged vegetation could be detected because it reflected less infrared light than healthy vegetation. In the experiment, aircraft flew over 210 test areas, each 1 mile by 8 miles, scattered along 3600 miles of flight lines in seven corn belt states, collecting data at intervals of approximately two weeks. Each aircraft carried a variety of sensors, including a multispectral scanner that simulated equipment planned for Landsat.[1] Supporting research included low-altitude aircraft and ground surveys. Methods for data

Figure 13
President Richard M. Nixon and Agricultural Secretary Clifford Hardin are briefed on use of the RB-57 high-altitude aircraft in the Corn Blight Watch Experiment by Air Force Major Francis X. McCabe and Dr. Archibald B. Park (far right), June 24, 1971. (NASA)

analysis were developed by the University of Michigan group that became the Environmental Research Institute of Michigan and by the Laboratory for Application of Remote Sensing at Purdue. The Corn Blight Watch successfully detected corn blight in the early stages of infection, when the blight could be checked by pesticides in time to save the crop.

The Corn Blight Watch experiment was successful, but it did not produce the results intended by the space agency. It led to favorable publicity for NASA and the operational use of commercial low-altitude aerial photography for the early detection of corn blight, not increased interest in similar uses of satellite data.[2] The successful use of remote sensing to monitor crop health depended on detecting the infection in individual fields or portions of fields, to catch the disease before it spread. Therefore, coarse-resolution data of the type expected from Landsat were less suitable than aerial photography. In order to find agricultural uses for earth resources satellite data, NASA had to turn to research on yield prediction instead of disease detection.

Early experiments showed that using a satellite to measure the areas planted with various crops had intrinsic difficulties. Using simulated Landsat data provided by an experiment carried on *Apollo 9* in 1969, scientists achieved crop identification accuracies varying from 30 to 90 percent in experiments examining the Imperial Valley of California and the area around Phoenix, Arizona.[3] Robert N. Colwell of the University of California, one of the pioneers of remote sensing, concluded from this study that better accuracy would require repetitive coverage as the crops changed their appearance during the growing season. On the basis of early studies, many specialists at the Department of Agriculture had serious doubts as to whether the coarse resolution planned for Landsat could provide useful estimates of crop acreage.[4]

Research on identification of crops revealed difficulties not only with coarse resolution but also in computer programing. Staff at Purdue's Laboratory for the Application of Remote Sensing made slow progress on developing a computer program to identify crops automatically from remote-sensing data (figure 14), initially using data from the corn blight experiment. As early as 1972 another group of researchers at the Environmental Research Laboratory of Michigan had recognized and was working on one of the key problem areas for computer analysis, called *signature extension*. The essence of the problem was that a computer program used to tell

Agricultural Applications of Landsat Data 149

Figure 14
Enlarged Landsat image showing fields in California's San Joaquin Valley (top). Computer classification of the crops in those fields (bottom). (NASA, 1976)

wheat from corn in one place would not necessarily work a few hundred miles away or on pictures taken a few days later. Substantial variations in appearance resulted from changes in soil, water supply, variety of the crop chosen by the farmer, the transparency of the atmosphere, and the growth of the crop.[5]

Despite these potential difficulties, NASA officials planned a major experiment on crop-yield forecasting because the space agency needed support for Landsat from the Department of Agriculture. Initial leadership within the space agency came from Robert B. MacDonald, who became the Landsat Chief Scientist at Johnson Space Flight Center in 1971. As leader of the Corn Blight Watch team at Purdue, he had become aware of a lack of interest in Landsat among scientists and resource managers in the operational bureaus of the Department of Agriculture, who NASA had hoped might profitably use Landsat data.[6] NASA particularly needed enthusiasm for Landsat from Agriculture's Statistical Reporting Service, which used an expensive system based on ground surveys by field agents to predict U.S. harvests. NASA policymakers believed that this system might profitably be replaced by a system using satellite data.

NASA needed a case where Landsat data replaced existing methods and saved money so that the agency could justify the Landsat project to the Bureau of the Budget. At a meeting sponsored by the Office of Science and Technology—the staff of the Science Advisor to the President—Theodore Byerly of the Department of Agriculture stated that the cost-benefit ratio predicted for Landsat "was not high enough to make the earth resources program a high priority activity within Agriculture."[7] Before NASA could hope to obtain adequate funding for earth resources satellites, the space agency needed more enthusiasm from the user agency that, according to NASA cost-benefit studies, had the largest potential gain. More participation by the Department of Agriculture would also help NASA balance the tendency of the Department of the Interior to try to take over the Landsat program. NASA hoped that a successful demonstration project would create the desired enthusiasm at Agriculture.

In 1973 political events made funding for an agricultural survey project easy to obtain. National attention was turned to crop forecasting by the grain export situation of 1972, when inaccurate predictions of Soviet harvests cost American farmers the high profits they could have earned if they had had prior knowledge of

Soviet shortages. In July 1973 NASA officials were questioned in Congress about why the agency had failed to do more to predict foreign yields.[8] In reaction NASA Headquarters responded favorably when Robert MacDonald suggested a large-scale agricultural experiment.[9]

A worldwide agricultural survey demonstration project was discussed at a meeting called by NASA Associate Deputy Administrator Willis H. Shapley on August 19, 1973. The outlines of the project emerged quickly: Reduce the scope of the problem by starting with only wheat, focus on measuring the area planted, and develop sophisticated techniques to perform the analysis quickly, since operational information would have to be timely. The meeting ended with a decision that the Earth Resources Program Office at Johnson Space Center should prepare a detailed plan that included "the proper utilization of the people from other appropriate agencies."[10] By October 1973 project management planning for what was then called the Large Area Crop Inventory Project had started at Johnson, and the Office of Management and Budget (formerly the Bureau of the Budget) was considering the project for approval.[11]

Organization and Results

NASA had originated the LACIE project for internal bureaucratic reasons, not because of clear user demand. Political pressures only increased over the course of the project. In addition, NASA managers chose technology for the experiment to fit NASA's technological interests, which did not always match the needs of the potential users. The tensions that resulted from this approach shaped the development of the project.

While scientists at the Johnson Space Center started work on the project in 1983, interagency tensions delayed the official beginning of LACIE until late 1974. Final approval for the project from NASA Headquarters was delayed by difficulties in negotiating an interagency agreement.[12] Secretary of Agriculture Earl Butz finally approved the LACIE project in the form of an interagency Memorandum of Understanding in late September 1974. NASA representatives signed quickly, but the National Oceanic and Atmospheric Administration, which was to develop the models of the effect of weather on yield, withheld its approval until late October because of funding difficulties.[13] Because of these delays and because Agriculture was slow to establish the administrative mechanism it needed to participate in LACIE, many decisions were made without the department's input.[14] Department of Agriculture scientists eventually developed a set of requirements for LACIE, but these requirements came too late to have much influence. Because the project was already underway, NASA managers decided that the Department of Agriculture's requirements would "be satisfied within LACIE only where practical."[15]

The conflicting definitions of the project involved approaches to research, not just organization. In one NASA meeting the discussion turned to attitudes toward LACIE at the Department of Agriculture. In particular, managers at Agriculture apparently had reservations about the project because of NASA's reputation for pursuing technological sophistication for its own sake. The first draft of the minutes of the meeting quoted NASA's Associate Administrator for Applications, Charles W. Mathews, responding to this issue by saying: "The project should be so structured as to avoid the tendency for a technical 'overkill.' We should strive to minimize the resources employed but yet assure the best potential for operational readiness."[16] A few weeks later, a memo corrected the meeting minutes: "In fact, it was stated that in effect we should, for such an advanced project, follow the NASA tradition and encourage some technical 'overkill.'"[17] In other words, NASA managers felt that the best way to test the potential of Landsat data for agricultural survey was to use the most sophisticated technology for analyzing that data.

While such an approach made sense for an agency whose mission was research and development, it increased costs and complexity for potential users. The Department of Agriculture group complained that the NASA people assigned to the project were mostly engineers left over from the Apollo program who had little knowledge of agriculture. That complaint involved both a sense that LACIE was a make-work project designed to meet the needs of NASA, not the Department of Agriculture, and also a divergence of approaches to system design. The agriculturists had little respect for the engineers' preference for developing self-contained models that would allow computers to make predictions unassisted by human judgment, if provided with huge amounts of data.[18]

The first phase of the newly approved experiment added technological difficulties to the project's problems. Phase One tested methods for predicting wheat yields in the Great Plains region of the United States, for which ground truth and detailed yield information were available. Signature extension was identified as the key problem as early as May 1974, and a "tiger team...was established to solve this problem."[19] In August 1975 the project team abandoned its attempt to use signature extension in the proto-operational portion of the project because "successful cases of signature extensions have occurred in only 20% of the attempts and in a random manner."[20] Despite this failure, the first yield estimates produced by the LACIE methodology using 1973–74 data were very successful; the results were within 3 percent of the official Department of Agriculture statistics. However, the larger-scale phase-one test was disappointing. By midsummer the LACIE estimates were running 40 percent above the official Department of Agriculture predictions for Kansas.[21] The final result for the year showed improvement, with an error (variance) of only 7 percent in the estimate of the area planted in wheat, but the predictions of the yield per acre in that area were not as satisfactory.[22]

During the first phase of LACIE, NASA sponsored a variety of research projects to improve estimates of the area planted in wheat. The primary LACIE methodology was based on sampling. Skilled photo-interpreters, assisted by a computer, differentiated wheat from other crops in selected small areas, and the computer then generalized to find the total area of wheat. The sample areas had to be chosen to accurately represent the total region of interest, and this selection depended heavily on sophisticated statistics.[23] In addition, experiments had shown that the best results could be achieved by looking at crops in particular stages of development. A schedule of stages for each region, called a *crop calendar*, had to be developed. The whole process of analysis had to be adjusted to allow for variations in the crop calendar with weather conditions in a particular year. A scientist at Kansas State University, Arlin Feyerherm, developed an improved crop calendar for wheat. However, his research led him to doubt the LACIE methodology, because he found that crop yields could vary a good deal between nearby locations.[24]

For LACIE to achieve its accuracy goals, research was needed not only to improve estimates of the area planted in wheat but also to improve predictions of the yield from that area. A team from the National Oceanic and Atmospheric Administration worked on a model for predicting the yield of wheat per acre on the basis of weather. This was difficult because the effect of precipitation on a crop depended critically on the timing of the rainfall relative to the crop's development. The goal for LACIE, decided on early in the project, was 90 percent accuracy in predicting wheat yields, nine years out of ten. This meant that the sampling error had to be kept to 2 percent, the error in the acreage estimation to 7 percent, and the error in the yield estimate to 7 percent (the errors were not additive).[25]

The second phase of LACIE was conducted under more political pressure, and the participants had to deal with a whole new set of problems. The Office of Management and Budget required that NASA justify the cost of *Landsat 4* by its contributions to an operational crop forecasting system.[26] According to rumors in 1975, NASA management had decided not to launch the spacecraft unless phase two of LACIE was successful in the accurate prediction of crop yields for other countries.[27] This was a more difficult task than predicting U.S. yields because much less ground information was available either about varieties of crops and soil conditions or about the progress of the crop in the year being studied. The international part of the LACIE project also had to be handled with special care because of international sensitivity to what might be perceived as spying and because of the potential economic value of yield predictions to investors playing the commodities market.

By reducing the scope of the project, NASA achieved success in the main objectives of phase two of LACIE. The original plan called for monitoring wheat in eight countries, but this number was gradually decreased. A feasibility study completed by Lockheed in early 1975 predicted that wheat areas could be identified and measured in the USSR but not in the People's Republic of China or in India because in those countries crops were planted in fields too small to be distinguished in Landsat images. Budget limitations and the lack of success with signature extension (with the result that more samples had to be examined by human analysts) meant that the project had to be scaled down. NASA decided to concentrate the primary phase-two effort on the United States and the USSR. The agency also joined with Canada in a cooperative project for predicting Canadian yields. The Canadian project had little success because in Canada wheat was often planted in narrow strips alternating with fallow land or other crops. This hopelessly confused analyses limited to coarse-resolution data such as those from Landsat, which did not show individual strips. Strip planting also caused problems in analyzing data of North Dakota. In addition, in colder climates farmers usually planted spring wheat instead of winter wheat, and it turned out to be nearly impossible to differentiate spring wheat from other spring small grains, such as rye.[28] In the end, LACIE scientists approached the problem of separating spring wheat from other small grains by first determining the total area of all spring grains and then calculating the area planted in wheat by simply using the ratio of crops planted in that region in the past.[29] Despite these difficulties, phase-two results for the USSR were good, and some U.S. results were excellent. However, predictions for Canada and the northern United States did not meet the performance standards of 90 percent accuracy.[30]

The LACIE phase-three analysis of the 1977 growing season succeeded in meeting the goals that the project had set. Again, the results for the U.S. Great Plains and the USSR met the accuracy criterion, with fewer person-hours of work than required the year before. Some progress was even made toward solving the problem of differentiating spring small grains.[31] LACIE successfully predicted poor USSR wheat harvests In 1977.[32] However, it was questioned at the time whether LACIE was capable of consistent accuracy or whether the good results were due to luck. The General Accounting Office attributed the accuracy for the USSR to offsetting errors. Canada had been dropped from the experiment, and there was considerable argument about the precise statistical meaning of 90 percent accuracy nine years out of ten.[33] Thus, the project ended with positive but mixed results.

Transition to Operational Use

The operational bureaus of the Department of Agriculture were not impressed with the results of LACIE and did not adopt the system for predicting yields developed in the project. The Department chose not to adopt the LACIE system because it had not been designed to suit the needs of agency and because of concerns about the costs of the LACIE system. Instead, one branch of the Department of Agriculture developed its own system based on a different approach to the use of Landsat data.

The Statistical Reporting Service, which had the operational responsibility for U.S. crop forecasts, showed little interest in the project and questioned the validity of the methodology.[34] Statistical Reporting Service personnel had an interest in maintaining their existing system, which created goodwill by hiring large numbers of people to collect data.[35] In addition, cost was clearly a central issue.[36] Despite pressure from the Office of Management and Budget, the LACIE project did not include the calculation of accurate estimates of operational costs to make possible a detailed cost-benefit analysis.[37] Because of the large amount of research and development work performed to support LACIE, and the many human analysts required to interpret the data, LACIE cost about $60 million and involved up to 400 people.[38] Without a more streamlined system, the benefits would certainly not exceed the costs. Even with streamlining, the technologically sophisticated system NASA had designed would always be expensive.

While the Statistical Reporting Service did not adopt the LACIE system, the results of the project at least increased Agriculture's interest in remote sensing.[39] Proponents had hoped that the LACIE project would result in an operational system that would be turned over to the Department of Agriculture. Instead, NASA, the Department of Agriculture, and the National Oceanic and Atmospheric Administration first set up a year of LACIE follow-on studies and then, in 1978, began a continuing experimental program called AgRISTARS (Agricultural and Resources Inventory Surveys through Aerospace Remote Sensing). The Department of Agriculture contributed half the funding for AgRISTARS, a significant increase over the 10 percent they had contributed of the resources required for LACIE.

A smaller branch of the Department of Agriculture, the Foreign Agricultural Service, showed more interest in using LACIE techniques, but it could not afford to take over the LACIE technology as a complete system.[40] Also, a system that achieved its accuracy goal for one crop in a few areas of the world where grain was planted in large fields met only a few of the needs of the Foreign Agricultural Service, which was responsible for predicting foreign harvests.

Instead, the Foreign Agricultural Service developed its own operational system for using Landsat data. Although based on general knowledge and basic technology from LACIE, the Crop Condition Assessment System introduced in 1979 depended on an entirely different approach invented by the Foreign Agricultural Service. LACIE methodology involved measuring the area planted in particular crops and then finding total production by multiplying area planted by yield per acre. The Crop Condition Assessment system, on the other hand, compared Landsat images from one year to the next to detect changes resulting from different amounts of land planted in various crops or from bad weather. Only changes were measured, not total area planted. To make comparisons possible, the system used a data base that divided the world into a fixed grid and stored weather information and digital Landsat data for points on the grid. An analyst using a small computer compared Landsat images and called up information from the data base. Because this system did not bother to calculate yields that showed no change from the previous year, a staff of twenty-eight could monitor crops all over the world and provide early warnings of bumper crops and bad harvests.[41] This system was innovative both in its simplified approach and in its use of Landsat data as part of global information systems.

NASA had lessons to learn from LACIE about technology transfer. A policy study of the Foreign Agricultural Service system stated that

there was no real transfer of technology from R&D to operational use during LACIE. LACIE/NASA considered the transfer accomplished when the demonstration project had been completed and success declared. There was very little attempt to help USDA make the transition from a large unwieldy R&D system to an efficient and affordable operation.[42]

Some Foreign Agricultural Service personnel suspected that NASA not only failed to plan effective technology transfer but also was particularly unhelpful to the Foreign Agricultural Service because its system rejected the LACIE methodology. Also, the Crop Condition Assessment system supplemented rather than replaced the previous sources of information. Therefore, the Foreign Agricultural Service system did not save any money that NASA could point to in justifying Landsat to the Office of Management and Budget.[43]

In the case of agricultural survey, the operational use of Landsat data developed in spite of, rather than because of, NASA's technology transfer efforts. While NASA's LACIE project may have served to educate the Department of Agriculture about satellite imagery and Landsat, the specific technology developed was not adopted because it was politically inexpedient for one branch of the Department of Agriculture and inappropriate in scale for the other. NASA had to deal with the operational branches of the Department of Agriculture rather than with a group of enthusiastic research scientists, as it had in the EROS Data Center. NASA specialists knew how to design sophisticated technology to get a job done. However, they did not always succeed in their efforts to design technology appropriate for small-scale users or involve users in the design process. The space agency also lacked skill in overcoming the numerous political and organizational barriers that lay in the path of the new Landsat techniques.

13

Technology Transfer and State and Local Governments

The negotiation of uses of Landsat data took a rather different form with state and local governments and regional organizations. State governments lacked the bureaucratic position or the technological sophistication to negotiate compromises in the development of uses for Landsat data between their own interests and those of the space agency or to follow the course of the Foreign Agricultural Service and develop their own approach to using Landsat data. However, state officials and representatives came to provide an important political constituency for Landsat, and NASA eventually had to encourage their participation. NASA managers conducted this outreach on a technology transfer model, but as that model became more sophisticated, it increasingly approximated a negotiation between equal interest groups. The development of NASA's technology transfer program thus showed how space agency managers came to understand the role of the users in defining how Landsat data would be used.

After 1973 NASA officials directed the largest part of their efforts to encourage the use of Landsat data toward state governments and local and regional resource management agencies. This emphasis resulted both from an evaluation of where information needs were greatest and from political considerations. NASA had traditionally garnered support in Congress by careful geographical distribution of grants and facilities, and support for Landsat from state governments would clearly be politically useful to the agency. However, state and local users had many of the same reservations about the usefulness of Landsat as the Department of Agriculture and had fewer resources for redefining the use of Landsat data to suit their needs.

State governments needed new information sources for resource management, but states also lacked both the expertise and interests that would have allowed a smooth adoption of the use of Landsat data. The problem involved both the techniques NASA managers used for technology transfer and the nature of the technology or techniques for data use that the space agency wanted to transfer. It took NASA a long time to develop effective methods for encouraging state agencies to use Landsat data and for transferring advanced techniques for using the data, and even then resource managers found the applications developed by the space agency not always well matched to user needs. When NASA tried to transfer techniques for using Landsat data that were too expensive and too complex, state governments simply failed to adopt the new techniques. The states usually lacked the technological resources to develop their own applications better suited to their needs, although occasionally state and regional organizations followed that course.

Development of State Involvement in Landsat

Initially, the EROS program of the Department of the Interior took a more active role than NASA in encouraging state and local interest in Landsat. The Department of the Interior had always worked closely with state governments on resource management. In 1969 the Geological Survey awarded a grant for a one-year demonstration project to investigate how the state of Washington might use Landsat data when that data became available. The principal investigator was David W. Jamison, a research photogrammetrist at that state's Department of Natural Resources. Only simulated Landsat data were available at the time and the data were not entirely satisfactory, but the Washington study concluded that Landsat data would be useful to agencies at the state level.[1] The EROS Data Center provided training for interested state users during the early 1970s, but funds were not available for the center to conduct any large-scale systematic outreach program or to develop major new systems for state governments to use.

NASA began working with state governments on some small data-use projects during the same period. A November 1970 letter from Acting Administrator George M. Low called for the NASA centers "to apply multispectral surveys to pressing problems in neighboring regions" in order to develop specifications and methodologies for using remotely sensed data.[2] Such outreach was already well established at the Johnson Space Center, where the earth resources aircraft program had included state users from the beginning. The Johnson Space Center also supplied funding and personnel to an Earth Resources Laboratory established at NASA's Mississippi Test Facility in southwestern Mississippi in 1970. The leaders of NASA's centers realized that providing technical assistance to local authorities would benefit the centers in the long run (figure 15). By early 1972 the Johnson Space Center was cooperating with, among others, the Texas Forest Service, the General Land Office of Texas, the Texas Parks and Wildlife Department, the Texas Education Agency, the Houston City Health Department, and the Texas Water Development Board.[3] Many of these cooperative activities involved simple uses of remote-sensing data, which could easily be adopted by the state agencies. Although established to test sensors for possible use on satellites, the aircraft program at the Johnson Space Center unintentionally encouraged the operational use of aircraft remote sensing. The director of the Texas

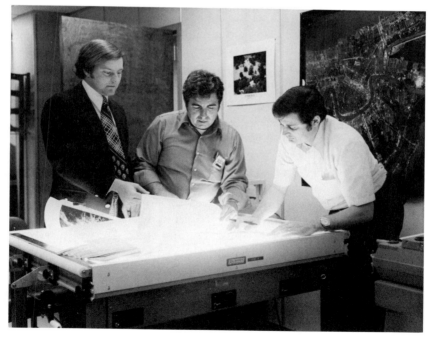

Figure 15
Louisiana state officials examine the use of Landsat data to delineate flooding by the Mississippi river. They are using facilities at the Department of the Interior's EROS Applications Assistance Facility at the NASA National Space Technology Laboratories, Bay St. Louis, Mississippi. (NASA, 1975)

Forest Service wrote to Johnson director Christopher C. Kraft, Jr., in September 1972 to thank the center for technical assistance that led the service to acquire its own aircraft and cameras.[4] In that case the user agency did succeed in redefining the technology to suit its needs.

During this early period, however, managers at NASA headquarters did not consider state governments a major target of the Landsat dissemination effort. A February 1967 memo showed not only the low priority given to state government users but also unrealistic expectations of technology diffusion:

To date, state level participation has not been encouraged although inquiries from state institutions are numerous... Such participation should normally require nothing more on the part of NASA than the supplying of copies of the imagery and related data with the request that the results of the investigation be made known to NASA in exchange.[5]

As late as 1973 NASA leaders resisted taking a more active role promoting Landsat data on the state level on the grounds that "for the present, at least, it is considered that communication with the ultimate users can best be promoted through the established channels of the various agencies involved."[6]

However, NASA sought support on Congress for Landsat on the grounds of its widespread usefulness, and the space agency soon faced questions from Congress about why, if the project was so valuable, state governments did not make wider use of Landsat data. In October 1972 a supporter of Landsat, Senator Clinton P. Anderson of New Mexico, wrote to NASA Administrator James C. Fletcher, describing awareness of Landsat at a meeting of state officials. Anderson wrote, "While NASA has done a fair job in setting up sources to which users could write for 'imagery,' the locations and capabilities of these centers were not well known."[7] NASA efforts began to increase in 1973 with a program of visits by space agency officials to state governments to promote Landsat data use. However, a survey done by the Johnson Space Center of "packaged techniques that could be of use to state agencies" found many "that looked promising, but few were ready for transfer."[8] This lack of easily adopted techniques, combined with widespread publicity about Landsat's usefulness, left some state government officials and representatives with the feeling that NASA was claiming much more than Landsat data could deliver.

Technology Transfer Policy

Even before Landsat, NASA had developed formal mechanisms for technology transfer to try to promote the spinoff of all kinds of aerospace technology, from Teflon to medical monitoring devices. NASA included finding new users for Landsat under the rubric of technology transfer, even though what was being transferred was not hardware but techniques for using new kinds of data. In 1971 Samuel I. Doctors claimed in a study of *The NASA Technology Transfer Program* that NASA's efforts to share new technology were second only to those of the Department of Agriculture. NASA's Technology Utilization Program, begun in 1962, initially emphasized providing literature describing new technology to potential users. When this bore little fruit, NASA established Regional Dissemination Centers at various universities to take a more active role in transfer. Even these centers were criticized for serving primarily as technical libraries. One study concluded that little transfer occurred because the dissemination of technical literature was not, by itself, an effective way of promoting the adoption of specific items of technology.[9]

By the mid-1970s space agency personnel had learned that in most cases providing information was not sufficient to ensure technology transfer; each new technology or technique had to be matched to existing problems and actively promoted before it would be widely adopted. An evaluation of NASA's technology transfer program in 1975 noted that

the graduation from Technical Briefs (shotgun dissemination) to Technology Utilization Surveys and Reports (packed-down technology) to Regional Research Applications Centers (where innovation is advocated) to teams of secondary inventors (who match and adapt space technology creatively to specific medical problems) is a continuing series of experiments to find the most effective way to diffuse technology by a nontrusted, nonintegrated agency.[10]

NASA was a nontrusted agency in the sense that potential users tended to fear that any technology coming from the space program would be too sophisticated to be practical or easily adopted. As the LACIE case showed, this lack of trust in NASA reflected a real difference in technological style. The 1975 study pointed to specific issues behind this lack of trust; successful techniques were often rejected by state governments because the supporting technologies they required, such as sophisticated computers, were "beyond the adopter's facilities or understanding."[11] NASA succeeded in spinning off technology to sophisticated private corporations, but the agency found it difficult to encourage the use of Landsat data by more conservative state governments.

Technology Transfer for Landsat

NASA officials gradually came to appreciate the need for a more effective program to promote the use of Landsat data and to realize that such a program required more attention to user needs. In 1973 NASA Administrator Fletcher wrote to the Office of Management and Budget:

> The first phase of ERTS [Landsat] investigations has been spent discovering and understanding the full extent of the informational content of space data; the next phase now needs to focus on making information extraction routine and on learning how the new information can be integrated into management and decision processes.[12]

The first set of Landsat investigations had produced few of the operational uses of the data that were becoming increasingly important in justifying Landsat politically. Under increasing pressure to prove the validity of optimistic cost-benefit estimates, NASA leaders developed a technology transfer program for Landsat data use.

Political pressure forced the space agency to change the emphasis and scope of its program to encourage the use of Landsat data. In 1972 ten NASA centers had participated in about 110 locally oriented experimental applications of Landsat data in 1972, and NASA had conducted 47 workshops and orientation sessions between 1968 and 1973, but most of these experiments and training sessions had involved scientists rather than potential operational users.[13] This situation had to change because NASA had to show that Landsat data had practical value to resource managers in order to obtain continued funding for the project (see chapter 7). The *Landsat 1* follow-on program of investigations (discussed in chapter 10) was more closely aimed at developing applications ready for transfer. Despite this, in December 1973 the Space Applications Board of the National Academy of Engineering criticized NASA efforts and called for the agency to undertake new approaches to technology transfer based on "an intelligent advocacy point of view."[14] In other words, NASA needed to push potential users to try the new data, not just tell them about it.

In 1974 the space agency began to investigate more systematically how to encourage the use of Landsat data. Albert T. Christenson, director of user affairs for NASA Headquarters's Office of Applications, established a User Communications Task Team in early 1974. This team organized a series of visits by NASA officials to state government organizations and planned future user-communications efforts.[15] Even this team started out emphasizing publicity rather than advocacy. The plan for state visits called for "establishing communications with higher levels of the respective State governments," not with the resource managers at the working level.[16] NASA leaders also tended to conclude that what was needed was more research and development, which was often true but also made it possible to avoid the question of persuading the users. This emphasis resulted in a series of large projects called *application system verification tests* (ASVT), which originally included LACIE (it was later reclassified as an independent project).

As chapter 12 has already suggested, the application system verification tests showed both technological difficulties and barriers to adoption of techniques for using remote-sensing data. One of the more successful application system verification tests used airborne radar to monitor ice thicknesses and distribution in the Great Lakes to serve as a guide for shipping and icebreakers. Ten percent of the ships in the area installed special equipment to receive the ice information that NASA collected and transmitted. The U.S. Steel Corporation supported the project by operating three transmitting stations in addition to NASA's four. The head of the project, Herman Mark, described the operation:

The winter of 1974–1975, due to vigorous exhortations on our part, arrived about three weeks late, giving us exactly the additional time required to design, procure, receive, build, install, test, redesign, modify, test, improve, rebuild, test, reinstall, flight test, repair, and finally put into operational use the ICEWARN system of our "interagency program."[17]

The result was that "for the very first time in Great Lakes history, the commercial shipping season did not shut down completely."[18] Luckily, the winter had been mild.

Despite the success of the ICEWARN experiment and its reliance on simpler airborne collection of data rather than Landsat, transfer still raised problems. Mark wrote that the Coast Guard showed a favorable attitude toward taking over the project operationally as planned, but Mark cautioned that

> they are extremely shy, however, as are all smaller and less well funded agencies. Though they appear to trust us at present and have already asked for our assistance in the North Atlantic, they will balk, and understandably, at the slightest sign or indication that we are patronizing them or acting in a way they do not understand.[19]

The Coast Guard did adopt the ICEWARN technology but did not progress to using Landsat data for similar purposes. Effective use of Landsat data for ice monitoring grew not out of NASA/Coast Guard cooperation but out of Canadian research efforts. Project ICEWARN illustrated the difficulty of transferring airborne remote sensing technology to the regional office of a federal agency, an easy task compared to transferring the use of Landsat data to most state agencies. Techniques for using Landsat data were more difficult to understand than aerial photography, and state agencies had less experience with sophisticated technology than the Coast Guard and were often more conservative.

Because of the limited success of the application system verification tests, NASA management reorganized the technology transfer program for Landsat in 1976. Managers at NASA Headquarters criticized existing large-scale tests as poorly coordinated. They also complained that many of the smaller projects sponsored by various NASA centers were primarily oriented toward creating regional goodwill rather than widespread use of Landsat data. Exactly what sort of transfer activity would be most productive turned out to be difficult to determine. The first reorganization plan, in April 1976, called for more specialization: a small number of regional Space Application Transfer Centers to assist potential users in utilizing Landsat data and to "help establish two-way communications between potential users and NASA."[20]

NASA had already experimented with this kind of facility at the Earth Resources Laboratory at the National Space Technology Laboratory (formerly the Mississippi Test Facility). In 1970 NASA had decided to locate programs other than rocket testing at the facility and invited other agencies to locate programs there as well. The Johnson Space Center proposed an extension of its earth resources program, to be located at the Mississippi facility and to concentrate on problems in that region. The director of this new Earth Resources Laboratory, Robert O. Piland, believed that NASA had to emphasize research on applications as well as research on sensors.[21]

The Earth Resources Laboratory at the National Space Technology Laboratory focused its efforts on local problems, working cooperatively with state, and local agencies, universities, and other federal agencies, including the Department of the Interior's EROS program. The Mississippi program included both developing new applications and training and assisting state and local users.[22] Initially, about half the laboratory's efforts were devoted to developing new large-scale applications through application system verification tests.[23] The Laboratory also provided intensive six-week courses in the use of Landsat data, emphasizing sophisticated techniques for using the data. By 1976 the training was limited to digital analysis, in contrast with the less technologically sophisticated approach stressing photo-interpretation used in courses at the EROS Data Center.

As the new technology transfer program developed, regional transfer centers became its focus. A study done in 1976 by Battelle Columbus Laboratories under a contract from NASA reinforced the regional approach. For the first time this study brought into the discussion theories of technology transfer. The authors of the report were most influenced by a book by rural sociologists Everett M. Rogers and F. Floyd Shoemaker on *Communication of Innovations: a Cross-Cultural Approach.*[24] Based on this understanding of technology transfer, the Battelle report recommended "a regional facility which brings information to potential users, demonstrates to them the advantages of the technology, and helps to persuade them to try it and adopt it."[25] The team at Battelle Columbus Laboratory discussed what was wrong with NASA's past technology transfer activities and concluded, using Rogers and Shoemaker's model of phases of invention:

Of the four stages—knowledge, persuasion, decision, and confirmation—past NASA activities have addressed primarily the first. Insofar as Landsat data are concerned, in the course of the Battelle survey we gained the impression that virtually all potential users had heard about Landsat, and know, at least in a general way, what its capabilities are. This suggests that NASA's past efforts have been fairly effective in completing the first stage. What is missing is a follow-through to the persuasion and decision phases. It is almost universally believed that this is most effectively done by face-to-face, one-on-one contact with individual users, usually with repeated contacts over a period of time.[26]

The authors of the study suggested that regional applications transfer centers, external to NASA, could take the necessary role of change agency. Such external centers could market the new sources of information to potential users more effectively than NASA, which was "widely distrusted among the non-aerospace U.S. industry."[27]

NASA management responded by expanding the regional applications transfer center program instead of creating external centers. The shape of the revised program was determined not by technology transfer theory but rather the aspiration of NASA Headquarters managers to exert tighter control over NASA field centers. The space agency initially planned just a few transfer programs, which could serve as centers of excellence, but by the summer of 1976 the plan called for a transfer center at each NASA center in order to deal with the jealousy and independence of the NASA field centers. Even then Christopher Kraft, director of Johnson, refused to participate, pointing out that his center already had successful and well-organized transfer programs. Kraft argued that appointing a space applications transfer officer, as requested by headquarters, would only disrupt the ongoing program.[28] Controversy also arose over the plan's specification that aircraft missions would be funded only when they supported a demonstration of the use of space-derived data.[29] Some centers had developed strong cooperative projects employing aircraft sensors to help local agencies, and the centers feared a loss of goodwill if those projects were terminated.[30]

Controversy within the agency and continuing criticism of the NASA technology transfer program by other government agencies led to another revision of the technology transfer plan in December 1976. This time NASA leaders replaced the proposed applications program at each center with three regional centers to be located at the Goddard Space Flight Center, the Earth Resources Laboratory, and the Ames Research Center in California.[31] NASA managers described this new plan as "responsive to the recommendations made to NASA by the National Conference of State Legislatures, the General Accounting office, and many state and local government users for a more focused and directed effort to provide technical assistance and training in Landsat applications."[32] Ben D. Padrick, in charge of the new user applications program at the Ames Research Center in California, quickly sent to NASA headquarters an expensive proposal designed to generate widespread user involvement. Recognizing that earlier attempts had failed, Padrick noted that his approach was "the only successful way of getting California involved on any reasonable time-scale because of their past experience with NASA."[33]

Although the plan for three regional centers provided more systematic technology transfer, it caused bureaucratic and political fallout. The field centers that were not included in the new plan complained about losing projects in Landsat data use. The change even elicited a protest from Senator Charles Mathias, Jr., of Maryland, who was concerned that Chesapeake Bay research being conducted at the Wallops Flight Center might lose funding. In reaction NASA Administrator Fletcher decided that the NASA centers not selected as regional applications centers should retain the capability "to adequately respond to user requests and carry on user-related applications activities."[34]

Despite these changes, criticism of NASA's technology transfer efforts continued. Although NASA officials had begun to realize the need to promote the new techniques, the space agency was not yet willing to give state and local users any influence in the Landsat project. The Intergovernmental Science, Engineering, and Technology Advisory Panel prepared a report in June 1978 giving state government perspectives on Landsat. The states had reached a level of involvement where they had started to try to take an active role, mostly by complaining. Governor Rueben Askew of Florida wrote, "This interesting new technology has not been used regularly by non-Federal interests and agencies because they had no input in the design capabilities of satellites."[35] The report claimed that the providers of space systems devised what they believed to be useful systems without consulting the prospective users and then oversold the new technology.[36] The report recommended an improved Landsat system and a systematic technology transfer program, including cooperative regional assistance centers.[37]

During the experimental phase of the Landsat program state programs tended to use it either as a small part of a large system or in ad hoc ways. Texas had a Natural Resources Information System, a data bank maintained and used by thirteen state agencies at an annual cost of about $300,000.[38] The Oregon Department of Agriculture analyzed Landsat images to find areas infested by the noxious weed tansy ragwort.[39] NASA even documented a case where Landsat and aircraft images were used by the Nebraska Remote Sensing Center to find a successful well site for a farmer named Galen Phihal, who had purchased an irrigation system and then had been unable to find a site for a well hole that would provide sufficient water.⁴⁰ Such uses were of limited value to NASA in arguing that an operational earth resources satellite system would be profitable.

By 1981 many states were using Landsat data, but effective user participation in decision making for the project was still a challenge for the future. Applications of Landsat data by state and local users came on all scales and levels of technological sophistication. A variety of NASA programs encouraged these applications but rarely in a successful, systematic way. Despite NASA's extensive technology transfer efforts, the space agency and state and local users did not have similar enough visions of how Landsat data should be used to enable widespread development of such uses.

14
Private Industry Users and the Early Debate over Privatization

Early in the Landsat project NASA leaders had to face the question of whether to turn Landsat or some part of the Landsat system over to private industry or whether earth resources satellites should be subsidized and operated by the government for the public good. The conflicting interests that centered on this issue were difficult to resolve in many government-sponsored programs for technological innovation. In the case of Landsat, relationships between private industry and government agencies were particularly complex. Private companies formed an important class of data users, and a small industry grew up to provide services and equipment for Landsat data analysis. Even those kinds of industry participation introduced new issues: industrial secrecy, government concerns about favoritism, and industry concerns about government competition. However, the greatest controversy arose over whether private companies would take the initiative and the risk of committing their own funds to owning and managing part or all of the earth resources satellite program. This chapter deals only with the question of privatizing the experimental Landsat system; the debate over privatizing an operational earth resources satellite system deserves a separate book.

Private companies could take a wide range of roles in an earth resources satellite project: They could be owners and operators of the system, users of the data, or developers and sellers of more sophisticated ways of processing and analyzing the data. The two existing models, communications and weather satellites, had set contradictory precedents (see chapter 2). ComSat and the established communications firms had taken the initiative away from NASA in the communications satellite field. On the other hand, weather data were defined as a government service rather than a commodity, and the operation of meteorological satellites remained the responsibility of government agencies during the 1970s. The Landsat cost-benefit analyses required by the Bureau of the Budget showed profits so high that skeptics wondered why private companies were not clamoring to launch this new class of satellite. Generally, NASA leaders discouraged privately controlled earth resources satellites for a variety of reasons, ranging from desire to keep control of the project to fear of misuse of the data. However, the agency had to encourage the private use of Landsat data to demonstrate the satellite's usefulness. Potential private industry participants also had conflicting interests: It benefited industry if government did the expensive research and development and then turned the results over to industry, yet industry did not want government agencies providing services that industry could sell at a profit.

Data Use

Most of the early attention NASA managers gave to private companies focused on their role as users of Landsat data. Early efforts at the space agency to encourage use of Landsat data in extractive industries quickly showed a whole series of new problems that grew out of dealing with private industry. The usual technology transfer issues continued: Questions arose over how much effort should be made to persuade industry to use Landsat data, and private companies probably redefined methods for using Landsat data rather than adopting those suggested by NASA scientists. However, the importance of these issues was unclear, because corporations refused to share information about how they used Landsat data. Proprietary information became the key issue that limited technology transfer.

The space agency gave the most attention to Landsat data use by oil companies. Research by *Landsat 1* principal investigators and work with Apollo and Gemini photographs suggested that Landsat data would be particularly valuable for oil exploration. Faults, salt domes, and other geological features associated with trapped petroleum could be identified more easily in Landsat pictures than from the ground or from low-altitude aircraft, because these features were often large and faint, better seen from a great distance (see figure 16).[1] Quick growth of sales and profits from this use would help justify the continuation of the Landsat project.

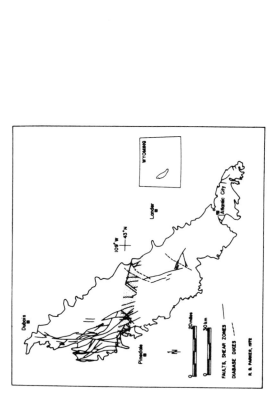

Figure 16

"Before" and "after" sketches of the geological structure in the Wind River region of Wyoming show the amount of new geological information resulting from Landsat. Knowledge of major geological fractures in this rather inaccessible region increased tenfold with the availability of Landsat images. Such information can be applied to mineral exploration. (NASA, 1973)

Soon after the launch of the first Landsat, the space agency sought to convince the oil companies of the potential of Landsat data. NASA Administrator James T. Fletcher discussed the results of geological studies of Landsat data at a meeting of the Houston Chamber of Commerce in August 1973. This elicited some interest, including a request from Conoco (Continental Oil Company) for a meeting with NASA earth resources personnel to obtain more information.[2] This request led to a one-day symposium hosted by the Johnson Space Center in November 1973 on the application of remote sensing to industry. The space agency invited twenty oil companies to send teams to hear about the investigations NASA had sponsored and to tour the Earth Resources Techniques Development Laboratory.[3] In fact, twenty-six companies sent representatives to the meeting, but they sent fewer scientists than Johnson Space Center managers had hoped for.[4] Conoco and Chevron, among others, expressed an interest in joint research projects with NASA.[5]

Just as with other technology transfer efforts, space agency managers found that users had doubts about the new technology. Some companies avoided using Landsat at all because they suspected that NASA was promising more than the satellite could deliver. These doubts were reinforced by a report by the Martin Marietta Corporation claiming that chemical changes in rock could be detected from Landsat data, a claim that even NASA reviewers described as "exaggerated beyond all belief."[6] Also, industry managers were interested in types of techniques for using Landsat data that were different than those NASA scientists wanted to develop. NASA scientists criticized oil companies for using Landsat images only as a substitute for aerial photographs and failing to develop techniques to use the data to their full potential. As a consequence, oil and mining companies complained that they needed a better satellite than Landsat, one that would produce images more like aerial photographs. Instead, NASA leaders wanted to encourage the companies to learn how to obtain the information they wanted by applying techniques more sophisticated than photo-interpretation to Landsat data.[7]

In addition to NASA's usual difficulties with technology transfer, problems quickly arose because of the proprietary nature of petroleum exploration research. Oil companies often refused to work with NASA in developing new techniques because the results would then have to be made public.[8] The issue of proprietary information also made it difficult even to evaluate industry use of Landsat data. The extractive industry kept its uses of Landsat data secret to protect both new methods of analysis and information about what regions were being studied. Some companies even ordered Landsat data through blinds to conceal their interest in particular areas.[9] The EROS Data Center considered information on customers' orders to be private and refused to release it.[10] Compounding the problem of cooperative research, NASA leaders worried about giving economic advantages to individual companies. To avoid such a situation, Deputy Administrator George M. Low suggested that NASA should exclude commercial firms from consideration as Landsat follow-on principal investigators.[11] Some NASA managers, in fact, believed that the best way to conduct research on the use of Landsat data in detecting minerals was in cooperation with a foreign country that had a government-run extractive industry.[12]

Despite these problems, the extractive industry put Landsat data to a wide range of uses fairly quickly. For example, by mid-1975 Chevron Oil claimed to be spending $50,000 to $100,000 per year on obtaining and interpreting Landsat data.[13] In particular, the company used the satellite data to make quick-and-dirty geological maps of large areas, a task that would have been extraordinarily expensive by any other method. Chevron also used Landsat images to prepare new charts warning of reefs and changes in channels in poorly charted coastlines.[14] Despite this apparent emphasis on using Landsat data as a replacement for maps in poorly mapped areas, American companies widely used Landsat data of the United States, where good maps were available. In late 1975 the EROS Data Center added up the five most recent orders of the ten largest industrial users and found that 72 percent of the 414 scenes purchased were of the United States.[15] This suggested that oil and mining companies used Landsat data not only to make maps but also to decide where to drill or dig.

Despite industry doubts and the slow development of advanced techniques, the use of Landsat data clearly brought substantial benefits to the American extractive industry. Company secrets and the tendency to use Landsat data in conjunction with many other exploratory techniques made rigorous cost-benefit studies impossible. However, the director of exploration of a lead mining company expressed a common opinion when he said that "Landsat as a prospecting tool is the most important development since the invention of the magnetometer in the 1940s."[16] In 1978 the president of the Earth Satellite Corporation, which provided data analysis services for other companies, estimated that at least $1 billion worth of oil had been found using Landsat data.[17] Even with adjustments for exaggeration that suggests significant benefits.

Landsat data also proved useful to other industries, although adoption was slower. In mid-1976 the EROS Data Center identified eight applications as operational or in testing for operational use in private industry. In addition to mineral and oil resource surveys, private companies were using Landsat data for site selection and geological impact studies for civil works projects, for finding likely sources of groundwater, for monitoring the use of water in irrigation, and for mapping soils and vegetation.[18] Despite these applications, there was a a sharp drop in the sale of Landsat data to private industry after 1975, with only a minor increase in 1979, as figure 12 shows. Part of this decrease may have been due to overenthusiastic buying in the first few years, part to the fact that mineral exploration applications did not require repetitive coverage of the same area. Despite these explanations, NASA could only be disappointed that private industry's use of Landsat did not show more rapid growth.

Data Analysis Services and Equipment

NASA's technology transfer efforts, particularly with private users, had to be established with great care to avoid competition with another type of private industry involvement—companies that provided data analysis services and equipment. NASA and the EROS Data Center processed and disseminated the basic data provided by earth resources satellites, but a private industry quickly developed to provide more sophisticated processing and analysis of the data and to sell equipment for such processing. By May 1978 there were thirty-seven companies providing processing services, twenty-three companies that sold image-processing hardware systems, and three that provided image-processing software systems.[19] These companies took a wide range of approaches to the use of Landsat data, but their financial interests tended to encourage a model of Landsat data use based on sophisticated information extraction and data processing that could be sold for a high price because of its scarcity. In some respects this definition of Landsat data use resembled NASA's, which prompted industry criticism of NASA efforts in the same direction as unfair competition.

This new value-added data processing industry produced specially processed data and interpreted data for the specific needs of various users. The users were businesses that were too small or lacked the appropriate technology to do their own data analysis or that occasionally needed more sophisticated processing or analysis than their in-house capabilities provided. For example, an agency or company that used Landsat data to make maps might pay extra for a finer-resolution image, whereas a company in the business of buying and selling grain might want to buy forecasts of harvests based on Landsat data rather than buy the computers and develop the expertise necessary to make its own forecasts. To supply the needed flexibility, some of the value-added data processing companies developed all-digital data processing systems that produced finer resolution than the standard product provided by the EROS Data Center. Computer analysis made it possible to alter the images' colors, enhance particular types of features, and even classify types of vegetation and land use.

Many of the earliest participants in the new industry were university laboratories and spinoff companies that already did research on aircraft remote sensing and the potential of Landsat data. The Institute for Science and Technology at Willow Run Laboratory of the University of Michigan became a private nonprofit corporation, the Environmental Research Institute of Michigan, in 1972. Under the sponsorship of the National Science Foundation and NASA, Willow Run had conducted courses on remote sensing for U.S. and foreign scientists starting in 1966. The Institute continued this work after separating from the university. Other projects, such as monitoring thermal effluents from power plants for the state of Michigan easily crossed the line from sponsored research to sale of a specialized service.[20] The Environmental Research Institute also worked extensively with foreign countries, providing planning and training to help countries set up their own remote-sensing agencies.[21] Other laboratories took on the same kind of projects without separating from their universities. By 1977 university laboratories providing processing and interpretation services included the Environmental Remote Sensing Applications Laboratory at Oregon State, the Office of Remote Sensing of Earth Resources at Pennsylvania State, the Laboratory for the Application of Remote Sensing at Purdue, the Remote Sensing Laboratory at Stanford, and four others.[22]

In 1970 a former NASA manager created the first profit-making company to provide special processing and analysis services for Landsat data. Robert Porter resigned from his position as head of NASA's earth resources program in 1970 to found the Earth Satellite Company, known as Earthsat. By July 1973, the end of the first year after the launch of *Landsat 1*, Earthsat had completed more than $1.8 million worth of business and had a backlog of another $2.1 million. By June 1977 it had achieved about $22 million in sales to forty-six federal and state agencies, fifty-six private companies, and twenty-four foreign governments.[23] In addition to doing research for NASA, the company cooperated in research investigations with foreign countries and private firms and provided training courses. Earthsat helped pioneer digital processing of Landsat data, using new processing methods to produce expensive, enhanced Landsat images that provided much more detail than the images sold by the EROS Data Center. The company also sold the results of its data analyses, particularly forecasts of crop yields. These provided speculators not only with forecasts of actual yields that Earthsat claimed were more accurate than those of the Department of Agriculture but also with predictions of what the Department of Agriculture forecast would be.[24] In addition to providing data analysis services, Earthsat developed a sizable business assisting companies and agencies in developing their own facilities to analyze Landsat data.[25]

The leaders of the private companies that provided special processing and analysis services to Landsat data users repeatedly complained that the government kept competing with the services they provided. In particular, they accused the EROS Data Center in Sioux Falls of unfair competition when it provided training programs, assistance with analysis, and specially processed data. In testimony to Congress in June 1977 Porter complained that his company, Earthsat, had developed a capability to enhance the Data Center's standard products, only to have the standard products change:

Very recently, the Department of Interior has announced a new service program which very simply provides enhanced digital imagery. Speaking from a very parochial viewpoint, we have lost directly very substantial sales from private industry where we have a capability which is at least as good as that capability now being provided by Sioux Falls, and which we had in advance, and which is an example of a capability that private industry could have supplied. There was no need for Sioux Falls to make that extra step.[26]

Porter asked that the government define its role because "industry, particularly a new and small industry, cannot rationally make long term capital investments when it perceives that the government is and may continue to compete with it."[27] Similar complaints came from other companies.[28] In these cases the participation of private industry tended to discourage government innovation.

Companies that analyzed Landsat data, either to sell a service or for their own use, provided a market for another related industry—the manufacture of special equipment for data processing and analysis. Some entrepreneurs believed strongly enough that Landsat data would be widely used to take the risk of developing new products for the users. Some of the hardware systems sold for processing and analyzing Landsat data had originally been developed for processing aircraft photographs and perhaps also for processing data from classified reconnaissance satellites, but a significant percentage were developed specifically for Landsat data.

General Electric produced the first self-contained system for manipulating and analyzing Landsat images. As the prime Landsat contractor, GE had developed the basic data processing system at the Goddard Space Flight Center that provided preliminary processing for all Landsat data. Using this expertise, GE developed a relatively inexpensive special-processing system, called the Image 100. This system took a Landsat data tape produced by the basic system at Goddard and created an image on a television screen. The user could then manipulate the image by defining colors or changing the contrast to produce an image that classified land use into chosen categories or showed only one type of feature, such as water. The Image 100 consisted of a minicomputer (a PDP-11) with special components (hard-wired algorithms) to perform complex calculations for classification and geometric transformation and the scanners, displays, and printers needed to put data into the system and produce the results. GE produced the first Image 100s in the spring of 1974, and the Johnson Space Center sought approval in May of that year to purchase one for approximately $382,000.[29] The EROS Data Center also installed an Image 100 and in 1976 reported that it was the most popular of the three special-processing systems available at the Data Center.[30]

As with the main data processing system for Landsat, General Electric had chosen a fairly low technology, reasonably priced approach. Clearly, this better suited the needs of many users than the more usual emphasis on the most sophisticated processing. However, while the Image 100 was a success, it never came to dominate the market, which suggested that the increased information from more sophisticated analysis may have been more important than users initially realized.

Private Ownership

Private companies could use Landsat data and provide services and equipment for other users, but the suggestion that private companies might build and operate their own earth resources satellites raised a variety of problems for the space agency. Early in the project, industry had expressed some interest in operating the entire satellite system, but different conceptions of how the system should be used blocked development in that direction.

Even before the first Landsat launch, General Electric, the spacecraft contractor, had inquired about the possibility of selling additional satellites to private industry and foreign countries. A company or country might wish to buy its own satellite instead of buying data from NASA in order to obtain data for private use that would not be available to competitors or enemies. Arnold W. Frutkin, NASA's assistant administrator for international affairs, replied negatively to the GE proposal. He explained that an earth resources satellite providing proprietary data clearly violated policy set by President Nixon in a statement to the United Nations.[31] This prohibition of proprietary data killed the idea of private satellites; customers who would pay a high price for the exclusive use of Landsat data would not be interested if it were available to their competitors as well. The government prohibited proprietary use for reasons related to international relations and privacy issues. Countries and individuals would object to the collection of data about their land if they did not have access to that data and if the data were sold without their consent. Therefore, NASA made a commitment that all earth resources satellite data acquired with unclassified U.S. technology would be available to everyone.

The question of private ownership of earth resources satellites returned in the mid-1970s, when Landsat had proven itself at least a qualified success. Some policymakers suggested that the entire Landsat system should be turned over to private industry, with appropriate government regulation. NASA managers did not take this possibility seriously until the first official definition of an operational Landsat system in 1980 (discussed in chapter 16) but the early discussions illuminated industry's role.

A draft policy paper distributed by NASA headquarters in September 1976 divided the Landsat system into four segments: the satellite and data reception, basic data processing, data production and distribution, and customized information products and equipment. The 1976 paper gave NASA responsibility for the first two parts, the EROS Data Center for the third, and private industry the fourth. The policy paper also suggested that industry could take over the Data Center or its function, but its authors worried about the foreign-relations impact of private data distribution. The paper suggested that the investment and risks involved in an earth resources satellite system were so high that no company was likely to take on Landsat data distribution, much less the first two parts of the system, without a significant government subsidy.[32]

Despite this conclusion, interest in a privately operated system continued. In 1978 Presidential Directive 37, which called for a major study of civilian space policy, encouraged "domestic commercial exploitation of space systems." However, even this directive specified that "all United States earth-oriented remote sensing satellites will require United States government authorization and supervision or regulation."[33] Privately controlled satellites, if allowed, would have to be regulated to prevent unfair exploitation by private firms, particularly of data of developing countries. Despite President Carter's interest in privatizing government functions, the many problems of private ownership of Landsat could not be solved quickly.

These early discussions of privatization had at their heart different definitions of who would use the data. Proprietary satellites or data would tend to result in scarce, expensive information. Because of the high price industries would pay for exclusive data, industry believed that type of system would make a profit, but it raised serious questions about unfair use of the data. A privately owned Landsat system required to sell data on a nondiscriminatory basis to everyone was fairer but less likely to be profitable. In order to make a profit, the company that owned the system would have to increase the price of data, squeezing out smaller users, and leading to an increasing tendency to encourage and support only the most profitable uses. In addition, whereas regulation might prevent misuse of data, third-world countries might still suspect misuse (see chapter 15). The third choice, a government-owned system modeled after the use of weather satellite data for the public good, implied an emphasis on encouraging the use of data by foreign and domestic government agencies and citizens' groups. However, if Landsat data had clear commercial potential, government ownership of the system put the government in the role of competing with private industry, a particularly unpopular idea in the late 1970s and 1980s.

The participation of private industry in various aspects of the Landsat project brought up questions that later underlay the controversy over an operational earth resources satellite system. Was the goal of Landsat to save the government money by providing better ways to fulfill existing responsibilities? Was it to boost a new industry that would sell and use earth resources information around the world for American profit? Or was it to benefit all people by making possible better management of natural resources and land use? Conflicting interests had complicated NASA's relationship with the user agencies; NASA's relationship with industry involved another set of conflicting goals involving fundamental questions about the balance of government services and private enterprise.

15

International Issues for Earth Resources Satellites

As early predictions had suggested, Landsat data were often more useful outside the United States than inside, particularly in the early years of the project; yet earth resources observation satellites raised international political issues that were not easily resolved. Landsat data could be particularly useful to developing countries, many of which were poorly mapped and explored, but these countries also had good reasons to worry about how information about them might be used by other nations. The international debate over earth observation satellites raised a wide range of issues and involved a wide range of actors. This study narrows the topic by continuing to focus on the interaction between NASA and the users of the data, touching only briefly on the broader debate in the United Nations and other international forums.[1]

Other countries participated in the Landsat system in a variety of ways. They debated the legal issues it raised, particularly issues of whether satellite information about a country could be sold without that country's permission. A wide range of countries made use of Landsat data for their own purposes, and a few had the interest and resources to build ground stations to receive data directly from the satellite. Finally, a few countries began, even before 1980, to develop their own earth resources satellites. International use of Landsat showed some of the same patterns of transfer and redefinition of technology discussed previously. However, international users brought an even wider range of expectations and capabilities than users within the United States, and their different definitions of what an earth resources satellite should be were particularly revealing, especially in cases where other countries had the resources to redesign significant parts of the system to suit their own needs.

Origins of the International Program

From the beginning of the remote-sensing program NASA sought to encourage other countries to use data from earth resources aircraft and spacecraft. The first unclassified pictures of the earth taken from space were taken by Mercury and Gemini astronauts in the early 1960s from orbits that stayed near the equator. Therefore, most of the photographs available for study showed countries other than the United States. To avoid any political issues and to obtain ground-truth information more easily, NASA developed a number of projects for analysis of pictures of other countries in cooperation with those countries. Space agency leaders also encouraged the development of experiments in foreign countries as part of the earth resources aircraft program. Such experiments had both scientific and political advantages. A 1965 memo from the Director of Manned Space Science at NASA Headquarters to NASA's Assistant Administrator for International Affairs called for aircraft studies of sites in Mexico. The memo argued that "the full range of ground conditions necessary to calibrate remote sensors is not available in the United States" and also that "involvement of foreign governments in this program would underscore NASA's commitment to the peaceful uses of space."[2]

In early Landsat planning user agency leaders sought to involve NASA and Landsat data in cooperative international projects the agencies had already established. The Department of the Interior, in particular, had a tradition of cooperative research on the resources of other countries. For example, the Geological Survey, under the auspices of the Agency for International Development, had studied in depth an iron mining area of 7000 square kilometers in Brazil called the Quadrilatero Ferrifero of Minas Gerais. The Survey wished to extend this work with aerial and satellite data collection, arguing that it would be easier to build on existing agreements for cooperation and that the extensive research already conducted would provide valuable ground truth.[3] Using Landsat data as part of existing projects instead of organizing new international efforts under NASA leadership had both practical advantages and public relations advantages for the user agencies.[4]

These competing interests resulted in a tendency toward elaborate experiments with many cooperating agencies. In 1968 NASA established an International Participation Program of remote sensing research from aircraft and selected sites in Brazil and Mexico for the first experiments. The Latin American project began in June with a discussion of test sites and sensors at a meeting among representatives from NASA, the participating foreign governments, the Department of Agriculture, the Geological Survey, and the Naval Oceanographic Office.[5] In the early stages of the project, foreign scientists were trained in the United States and teams of U.S. scientists visited Brazil and Mexico to help with planning. Later, flights were made by U.S. aircraft carrying NASA sensors over test sites in those countries, and U.S. and foreign scientists worked together to analyze the resulting data. This cooperative effort led to another joint project in 1972. Mexico and the United States used remote sensing to help eradicate the screwworm fly from Mexico and prevent reinfestation of the United States by this disease-carrying pest. Specialists used aerial photographs to identify "environmental factors conducive to the growth of insect populations," such as altitude, soil moisture, and the amount of vegetation.[6] If remote sensing could identify relatively small breeding areas, pesticides could be used much more effectively.

Thus, certain types of projects were established even before the first Landsat was launched, and scientists in Brazil and Mexico had gained some experience with remote sensing data that would help them prepare to make good use of Landsat data. Although important issues remained to be worked out, NASA had at least brought international users into the project at an early stage.

International Law

These successful aircraft programs answered technical questions about international use of Landsat data but did not resolve issues of international sensitivity of satellite observations.[7] In airborne remote-sensing experiments, NASA always obtained advance approval of the subject country, but the space agency planned to take civilian satellite pictures without advance permission. The United States had established a policy of free observation by reconnaissance satellites by fiat, but civilian satellites raised somewhat different issues because the resulting data would be sold to anyone and because an open program could not be ignored in order to save face.[8]

NASA's handling of weather satellite images and photographs taken by astronauts had set a precedent for unrestricted collection and distribution of coarse- and moderate-resolution imagery gathered by spacecraft. Even that precedent had not developed without dispute; in the early 1960s the Soviet Union had complained about being observed by Tiros meteorological satellites. That complaint had not become a serious issue because weather satellite data had such coarse resolution that it was very unlikely that anyone could extract any information of strategic or economic value.[9]

However, problems arose for finer-resolution data. Most observers assumed that data gathered from space that had fine enough resolution to show military secrets would not be appropriate for open distribution. Experience with weather satellite data did not solve the problem of distinguishing between pictures showing geography and resources, which might be considered public information like maps, and pictures with such fine resolution that they divulged military secrets. Almost everyone agreed that the images returned by the first three Landsats would not reveal any information of specific military importance, but even data about resources might be used by other countries or multinational corporations to guide unfair exploitation of the resources of developing countries. Also, some countries feared (correctly) that allowing distribution of moderate-resolution Landsat data would open the door to open distribution of finer-resolution data.

Debate developed only slowly. The 1968 U.N. Conference on the Peaceful Uses of Outer Space discussed earth resources satellites, and no countries stated any reservations about NASA's satellite plans. In a speech to the U.N. General Assembly in September 1969, President Nixon committed the United States to a policy that made NASA's satellite data available to anyone in the world. Initially, most countries saw this policy as beneficial, but some countries began to raise questions as Landsat came closer to reality. The U.S. government tried to avoid the issue for Landsat by promising nondiscriminatory distribution of the data and arguing that an experimental satellite did not set a precedent. Despite these efforts, those countries that favored legal restrictions on operational earth resources satellite systems suggested that even Landsat data should have only restricted distribution.

The debate reached its height in the year before the launch of the first Landsat. In a January 1972 memo NASA's Assistant Administrator for International Affairs, Arnold Frutkin, admitted that "international questions on the legality and proprieties of ERS [Earth Resources Survey] data collection and distribution are surfacing," but "thus far we have been successful in muting these issues."[10] In September 1971 Mexico made it clear to the United Nations that it expected that "no data would be collected over Mexican territory from the air or space without prior permission."[11] The Soviet Union raised a different question, since both the Soviet Union and the United States were already collecting data with reconnaissance satellites without permission. The Soviets claimed that nations have a sovereign right "'to dispose' of their own resources and 'to dispose' of information relating to their resources."[12] Thus, data should not be sold to a third party without the permission of the subject country. The United States hoped to counter all these different objections by making data available on a nondiscriminatory basis to all rather than by agreeing to legal restriction or U.N. sponsorship of data distribution from Landsat.[13]

The United Nations tried unsuccessfully to solve these problems by drafting international remote-sensing law rather than by treating each new class of satellites on a case-by-case basis. In December 1969 the U.N. General Assembly asked its Committee on Outer Space to study international cooperation in remote sensing. The Committee on Outer Space was slow to focus seriously on legal issues relating to remote sensing; a legal subcommittee began work in 1974. Since NASA had already begun open distribution of Landsat data at that point, the legal subcommittee confined its task to regulation of operational satellites, arguing that it was impractical to try to regulate experimental systems. The subcommittee reached agreement on a number of principles but not on the core issue of when one country could legally collect information from space about another country and disseminate that information to third parties.

Different countries had very different opinions about how civilian remote sensing from satellites should be restricted. The United States wanted no restrictions on data distribution from civilian satellites so that data from operational satellites similar to Landsat could be sold freely. The Soviet Union and France proposed a draft law that would have allowed open dissemination of data except those of fine resolution.[14] Data with resolution finer than a certain limit could not be sold to third parties and had to be provided to the state being observed. Some of the developing countries, particularly in South America, insisted that the permission of the state being observed be obtained before any data were collected or distributed.[15] Various individuals proposed an international agency to run earth resources satellites instead of trusting an individual country to handle data in a fair way.[16] With such different approaches it was not surprising that no agreement could be reached.

The restrictions under discussion did not apply to Landsat because the United Nations's goal was to define the law for operational systems, but Landsat did set a pattern. As the number of Landsat customers grew and the U.S. government postponed the initiation of an operational system (see chapter 15), Landsat became a de facto operational system that set a clear precedent for the future. In addition, many developing countries came to weigh the benefits of Landsat more heavily than the dangers. A trade magazine noted in 1978 that "the energies of Central and South American officials, which earlier were directed toward criticism in the United Nations of Landsat type operations or concern over neighboring countries' use of the data, now is being directed more toward obtaining funding to use the data to boost national economies."[17] Without a formal legal agreement the common law for remote sensing from space developed on the basis of a broadly held perception that Landsat brought more good than harm.

The United Nations reached no final agreement on international law for earth resources satellites, but concern over the issue gradually moderated. When the United States started to move toward an operational earth resources satellite system that would possibly be privately owned, rhetoric over international remote-sensing law heated up again. The strong concerns of the developing nations were brought up and rejected again in the U.N. UNISPACE '82 conference held in Vienna in August 1982 when a group of seventy-seven countries submitted a position paper that included the statement "sensed states should have timely and unhindered access on a priority basis . . . to all data and information obtained over their territories. Dissemination of such data and information derived from it to a third party should not be done without the prior consent of the sensed country."[18] However, by the early 1980s most developing countries were more concerned about preserving open and nondiscriminatory distribution of Landsat data, which they felt was threatened by the possibility of a privately owned system.[19]

Use of Landsat Data by Developing Countries

NASA leaders sought to avoid demands for restriction of remote-sensing satellites by proving the value of openly available satellite data. From early in the project, NASA distributed samples of Landsat data to foreign governments and supported a program of experiments proposed by scientists from other countries. There were only two restrictions on foreign proposals: The proposals had to be approved by an agency of that country's government, and foreign investigators received no funding from NASA, only free data.[20] These foreign investigations and the wide distribution of goodwill copies of Landsat images resulted in a quick positive response by foreign countries to the launch of *Landsat 1*.

Initial benefits came quickly for developing countries. The United States had been completely mapped at fine resolution before the launch of the first Landsat satellite, but many developing countries had not. These countries could benefit vastly by using Landsat data just to provide accurate maps for the first time (figure 17). For example, Brazil reported just a few months after the first launch that Landsat images had already resulted in the discovery of islands larger than 200 square kilometers and the correction of the location of Amazon tributaries on maps by as much as 20 kilometers.[21] The U.S. Embassy in Mali reported that "the U.S. government has gained a million dollars worth of Malian political mileage" from Landsat.[22] With the help of the international courses at the EROS Data Center (see chapter 11) and other training programs, such uses of Landsat by developing countries multiplied.

Landsat data could bring substantial benefits to developing countries, but it required substantial technological investment and expertise. Working with Landsat data required special expertise in photogrammetry that was rare in third-world countries. Landsat proved most beneficial to governments that were willing to develop that necessary expertise and had enough control over their own resources through monopoly or regulation to ensure that the information was used beneficially.[23]

Foreign Ground Stations

A few countries chose not just to buy data from the United States but to participate more ambitiously by building ground stations to receive data directly from the satellite. This participation involved

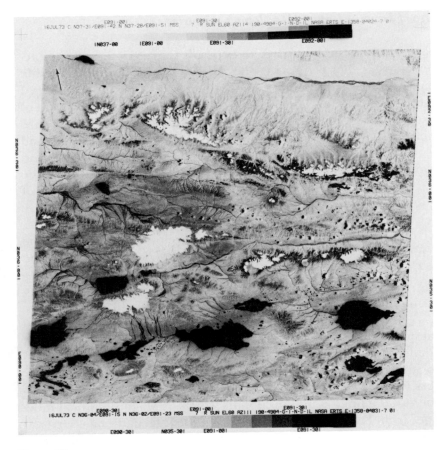

Figure 17
A high-plateau area dotted with lakes near the north-central edge of Tibet.
(NASA, 1973)

more international cooperation than NASA had originally planned for Landsat, but the Canadians and others persuaded the agency to provide them with data. The Canadians also designed a data processing system quite different from the U.S. model, redefining the technology to suit their needs.

The Canadian government developed its own ground station in time for the launch of the first Landsat. Aerial surveying with infrared and magnetic sensors sponsored by Canada's National Research Council had brought together a new scientific community, who became interested in NASA's plans for an earth resources satellite in about 1968. Led by Dr. Lawrence Morley of the Canadian Department of Energy and Mines, the remote sensing scientists convinced the Canadian government to support a proposal to found a Canadian Centre for Remote Sensing that would receive data directly from Landsat and distribute it within Canada.[24] The Canadian government then persuaded NASA to allow the Canadian Centre to receive data from the satellite and to make Canada an official participant in the project so that the Canadian team could receive all the technical specifications necessary for building their system.[25]

The Canadian Centre for Remote Sensing built a system of data processing for Landsat significantly different from the one developed at about the same time in the United States. Rather than buy a copy of the data processing system that GE was building at the Goddard Space Flight Center the Canadian group decided in early 1971 to develop their own system. They chose a different set of tradeoffs, and they hoped to learn from NASA's mistakes. Most significantly, the Canadian system added a feature the Goddard system lacked: extremely fast turnaround time for simple processing. A special quick-look system produced low-quality images for distribution within a few hours after the data were received from the satellite. These images proved particularly useful for monitoring ice cover that might interfere with shipping. Overall, the Canadian system cost only $6 million, although it produced slightly lower quality images than those provided by Goddard.

The Canadians did not avoid all the problems of NASA's data processing and distribution system for Landsat. The Canadian precision geometric correction system, although different from Goddard's, was no more satisfactory. Even producing standard products did not always go smoothly. One participant in the project remembered that at one point the electron beam recorder at the Canadian Centre for Remote Sensing took an average of eight hours to repair and then broke down, on the average, one hour later.[26] These similar problems in two fairly different systems suggested that many problems resulted from lack of adequate development of the technology, not poor design choices.

The Canadian Centre for Remote Sensing had not only some of the same technical difficulties as NASA but also some of the same problems with technology transfer. By the end of January 1973, six months after the launch of *Landsat 1*, Canada had distributed 50,000 copies of Landsat images, a respectable figure next to NASA's 218,000.[27] Because the Canadian system was smaller and less expensive than NASA's system, it was more flexible and could be changed more easily to suit the users. Despite this, technology transfer was still a problem. The leaders of the Canadian group found that persuading people to use the data operationally took longer than the five years the founders of the program had planned.[28]

A number of other countries followed the lead of the Canadians in establishing ground stations. Brazil and Italy were first, with the Brazilian station at Cuiaba starting up in May 1973 and the Italian one beginning two years later. At about this time NASA decided to charge a fee to the foreign stations, an option that had been included in the ground station agreements. NASA charged $200,000 per year to provide data to the foreign ground stations.[29] NASA managers hoped this would be a step toward financial self-sufficiency for Landsat. Even with the fee, many countries found ground stations worthwhile. By 1980 Canada had built a second station to cover its maritime provinces, and Sweden, Japan, India, Australia, Argentina, and South Africa had opened their own stations.[30] The cost of a ground station was not excessive; a 1978 estimate put the cost at about $6 million to build a complete ground station, with operating costs at about $1.7 million per year. A less ambitious station might cost about $1.9 million to build and $.75 million per year to run.[31] By 1980, these foreign ground stations played a significant role in international distribution, providing 20 percent of the total world sales of Landsat products (with the United States responsible for the remainder).[32]

Foreign Satellites

Those countries that wanted even more control over earth resources data considered launching their own satellites. Some of these provided only coarse resolution, but one, the French SPOT (Système Probatoire d'Observation de la Terre) satellite, improved significantly on both Landsat resolution and technology. Though discussed only briefly here, these later projects did reveal alternate choices of earth resources satellite technology and organization. The SPOT system also raised the question of why the United States, after pioneering earth resources satellite technology, was left behind in developing an improved commercial system.

Soviet interest in earth resources satellites showed the advantages and disadvantages of a system that did not separate civilian and military uses. The Soviet Union showed significant interest in Landsat, which suggested that their military reconnaissance program did not provide data as useful for resource studies or that they wanted to learn from U.S. approaches to interpreting Landsat data. As early as 1973 a U.S.-USSR Joint Working Group on the Natural Environment began discussing possible cooperative earth resources experiments.[33] Among other exchanges the Soviets provided ground truth from an agricultural test site in the Soviet Union to compare with Landsat data. Since the Soviet Union made no distinction between military and civilian programs, the Soviets probably used reconnaissance satellites for some earth resources studies. But they also developed satellites specifically for earth resources; as suggested in chapter 3, requirements were so different that military reconnaissance satellites were of only limited value for earth resources studies. The Soviet Union launched a series of Cosmos satellites equipped to photograph natural resources and in June 1980 an earth resources satellite called *Meteor* that carried a television camera.[34]

The interest expressed by several other foreign countries in launching their own earth resources satellites provided additional evidence of the value of Landsat data outside the United States. At least two interests led to independent satellite projects: desire for greater control over satellite data and perception of an opportunity for competing with Landsat by launching a better satellite. Developing countries generally could not hope to improve upon Landsat, but they could seek more data and more control over the data.

India launched a simple earth resources satellite carrying a television camera with a resolution of 1000 meters in February 1979 and made plans for a somewhat more sophisticated "semi-operational" system. Brazil started in the early 1980s to plan its own moderate-resolution earth resources satellite.

More developed countries saw a potential to compete with Landsat. The French led the way with a 1978 decision to develop the remote-sensing satellite called Système Probatoire d'Observation de la Terre (SPOT), first launched in 1986. The French aimed their earth-resources satellite at a commercial market from the start. For the purpose of marketing SPOT data, the French government helped set up and retained minority ownership of a company called SPOT Image. SPOT Image was established in 1982 and worked to develop a distribution system and solicit customers both for imagery and ground stations even before the launch of the first satellite.[35]

The French program involved improvements in technology as well as an emphasis on commercial use. SPOT carried more-sophisticated sensors than those on Landsat, with a resolution of ten to twenty meters and the capacity to produce stereo images useful for topographic maps.[36] The French plans leapfrogged American developments in sensor technology by starting out with multilinear array sensors. This new technology focused the light from the ground directly onto an array of tiny photoelectric cells, making it possible to record an image of a whole line of the scene at once instead of scanning each line. More sophisticated and potentially more reliable than the Landsat multispectral scanner, this design was even superior to the improved scanner, called the thematic mapper, that NASA launched on *Landsat 4* in July 1982. The French Space Agency plans called for a second SPOT satellite identical to the first and two improved satellites in the 1990s.

While France launched the first satellite to improve on Landsat, Japan had similar plans for a 1991 launch of an earth-sensing satellite sponsored by that country's ministry of international trade and industry. *Aerospace Daily* described the announcement: "Emphasizing the importance of worldwide resources data to Japan, which imports most of its energy supplies and raw materials, the ministry seeks nearly three times the analysis capability provided by Landsat data. It intends to furnish information from the Japanese system to developing nations."[37] Clearly, the Japanese space program planned to compete directly with U.S. satellites for sales to other countries, not just provide data for Japanese use.

French and Japanese satellite systems and plans raised concerns that despite its early lead, the U.S. earth resources satellite program might lose its market to foreign competition.[38] Clearly, the French program left NASA behind in earth resources satellite technology in the 1980s, particularly that technology most useful to the commercial market. The French redefined earth resources satellites to serve a commercial market while the U.S. program was still torn between different concepts of the definition of an earth resources satellite program, particularly what form of organization would be most appropriate and what kinds of uses should be encouraged.

Foreign countries not only brought new approaches to the earth resources satellite program not envisioned by NASA leaders but also provided examples where other countries had more success with their approaches than the U.S. space agency. More careful study of selected foreign cases would be necessary to determine whether Landsat technology transfer was significantly more successful in other countries; the survey provided here at least suggests a wide range of successful experiments. In part this suggests that infighting held back the U.S. program, but it also shows the power of the users of a technology to redefine it to fit their needs.

16
Early Steps toward an Operational System

Landsat supporters expected a successful experimental earth resources satellite to lead quickly to an improved operational satellite with the reliability and large-scale distribution system necessary for widespread use. An operational earth resources satellite system was in fact discussed even before the first Landsat launch, but the government made no official commitment to developing one until 1979. This left Landsat program managers in a trap: The experimental nature of the Landsat system made users reluctant to commit themselves to use the data operationally, yet the lack of widespread use bolstered the argument that Landsat was a failure and therefore should not be expanded into an operational system. In this situation progress toward an operational system was easily stalled by disagreements among the various interest groups, by NASA's desire for better technology, and by President Carter's plan for privatization. This chapter concentrates on the period before the 1979 commitment to an operational program, with only a brief survey of the transfer of that operational program to private enterprise.

Why did Landsat fail to lead to an operational system as quickly as expected? To some extent the reasons lay in a failure of the interested agencies to build the coalition necessary to gain firm support from the president and Congress.[1] But Landsat was not simply a policy failure because the lack of user enthusiasm for Landsat data was a solid ground for criticism by its political opponents. In part, NASA managers made the mistake of expecting user enthusiasm too early in the life cycle of the technology: Landsat enthusiasts predicted too much too soon. But, more important, the problems of conflicting definitions of the goals and shape of the Landsat system became doubly severe when the interested agencies tried to plan an operational earth resources satellite system. Most simply, each interested agency defended its own interests, but those kinds of conflicts are routine and can usually be dealt with effectively by the political system. In the case of Landsat, resolving those problems proved more difficult because those interests involved not just control of the problem but also dispute over the definition of the technological system. Different definitions would result in changes in the design and operation of the operational system and affect its value to other users.

NASA Reluctance to Plan an Operational System

NASA leaders had reasons to want to put off an operational earth resources satellite system into the indefinite future. The space agency wanted to prolong the experimental phase of the project, during which NASA would have more control and a larger share of the budget than the other interested agencies. More important, NASA leaders had a vision of technology development that emphasized technological innovation and thorough testing and assumed that the users could not adequately define in advance what kind of system they needed or what technology would best suit their needs.

NASA managers sought to preserve the agency's role in earth resources satellites by emphasizing the experimental nature of Landsat rather than its operational potential. An operational system would probably involve a smaller role and budget for the space agency, since NASA, as a research and development organization, controlled experimental satellite programs but not necessarily operational ones. In a response to a Bureau of the Budget request for a study of the usefulness of Landsat, a May 1970 NASA discussion paper stated:

Investigations based on *ERTS A & B* [*Landsat 1 & 2*] data are not "market oriented" in the sense that there are no operational commitments to established customer-service relationship envisaged. *ERTS A & B* data is not considered to be operational in that it would displace per se existing collection modes or fill gaps in operational requirements. The aim is to investigate the potentials of a new observational tool; while it is likely that some of the data will have immediate value in some of the areas of ongoing operational activity, that would be a bonus rather than a primary purpose. Any *ERTS A & B* commitment to operational activities would involve a far greater investment in processing facilities and a high risk of premature decisions in this area.[2]

NASA managers not only stressed the experimental nature of the first few Landsat satellites but also emphasized the need for a series of experimental satellites to provide a base of experience on which to design the best possible operational satellite.

Although NASA might not control an operational system, the leaders of the agency sought to control as fully as possible the development of the experimental system.[3] In a May 1971 draft of a NASA policy paper, David Williamson, a special assistant to the administrator, stated, "In the area of space applications it is NASA policy to provide service and support to the line institutions of the government charged with operational functions and to be guided by these line institutions as to the priorities among significant problems requiring technological solutions." However, in a statement on the next page Williamson severely limited his initial explanation of NASA's role as a service agency. He wrote:

Every effort will be made to obtain potential user input before development is started but, since development must frequently be started with incomplete user specifications, the principal criterion for deciding to go ahead may be the potential value of the technology involved rather than the immediate operational value of the particular system under consideration.[4]

In other words, NASA scientists and engineers would develop experimental systems without discussing them fully with the ultimate users, because the users often did not know what kinds of systems would be best. Falling into a pattern common in research agencies, NASA managers often wanted to pursue advanced technology for its own sake rather than for its ultimate benefits.

NASA leaders envisioned a lengthy series of experimental earth resources satellites, not early work on an operational satellite. For example, in 1972 the space agency began to study the possibility of an even more sophisticated experimental earth resources satellite system called Earth Observatory Satellite. Space agency managers wanted more research before the transition to an operational system. Considerable scientific interest justified testing advanced sensors such as the finer-resolution scanner called *thematic mapper* and various microwave sensors.[5] Before it was canceled, this project reached the definition phase, and its supporters were planning for design studies that would lead to a 1978 launch. However, NASA could not find support for another ambitious project for an experimental earth resources satellite, since the directions of technological development NASA scientists and engineers wanted to pursue did not match the interests of the user agencies.

Planning an Operational System

In the early 1970s, pressure from the user agencies forced NASA to start serious planning for a transition to an operational earth resources satellite. The different intentions of different agencies had been clear even in debates over the development of the experimental Landsat satellite, and they shaped discussion of an operational system from the start.

Even while stressing the experimental nature of Landsat, space agency leaders had started to argue for a large role for the agency in an eventual operation system. The precedents set by communications and weather satellites suggested that once a satellite program became operational, the organization that used the data would control the satellite system. NASA would then serve as a procurement and launch agent for the operational agency. For Landsat, however, NASA managers thought that their agency might retain the manager's role. A 1970 position paper suggested that Landsat would be valuable to such a wide spectrum of users that "it might be wiser to expand NASA's role to include service operations of the space segment, permitting each interested user to manage the transformation of data into information on his own time-scale and for his specific purpose."[6] That particular proposal probably represented a larger role than even NASA leaders expected the agency to actually end up with, but agency leaders usually find it wise to ask for more than they expect to get.

The interests of other agencies played a larger role than NASA concerns in shaping an operational earth resources satellite system. The Department of the Interior, and to a lesser extent the other user agencies and NASA, hoped from the start for an operational system. However, the Office of Management and Budget refused to allow funding for an operational earth resources satellite system into the budget, requiring that the interested agencies first prove that it would have a favorable cost-benefit ratio. Studies done before the first Landsat launch by Westinghouse, General Electric, Cornell University, and the Planning Research Corporation had estimated that the satellite would bring tens of millions to billions of dollars of annual benefits.[7] In practice, however, benefits accumulated only slowly and could not be reliably measured, and there developed no overwhelming enthusiasm among users for an operational system. To the extent that Landsat made possible better management of natural resources, it provided benefits to the nation and to future generations. However, these benefits did not appear in the form of savings to the government or of users willing to pay high prices for the data.

Pressure from the Department of the Interior countered the opposition of the Budget Bureau. After the launch of the first Landsat satellite in 1972 Department of the Interior leaders renewed demands for an operational earth resources satellite system under that agency's control, forcing NASA to organize serious joint planning for such a system. In July 1972 NASA proposed to study an operational earth resources survey system for launch between 1980 and 1985.[8] A few months later the Department of the Interior recommended an operational program that would begin in fiscal year 1974, a proposal that received strong statements of support but no effective action from NASA.[9] The two agencies combined their thinking in a joint study conducted at the Johnson Space Center, which some NASA participants hoped would lead to an operational system based on Landsat technology in fiscal year 1976.[10] Clearly, the two agencies (and even different levels within NASA) had divergent views on how much more research and development was needed to prepare for an operational system.

The 1972 joint NASA/Interior plan projected that as a first step NASA should develop a new satellite to test a system suitable for an operational satellite. NASA proposed a revised version of the Earth Observatory Satellite series as a proto-operational step. The first of these satellites would involve minor modifications of Landsat: an additional infrared band on the multispectral scanner and a more operational digital data processing system on the ground.[11] When the Office of Management and Budget denied funding for a proto-operational system, NASA renamed the improved satellite and flew it as *Landsat 3*. By treating the new satellite design as an improvement on an existing project rather than as a new start, NASA managers sidestepped the task of getting approval for a new project.[12] Similarly, when NASA developed a new sensor for earth resources observation, the thematic mapper, the agency chose to fly it on another Landsat (*Landsat 4*) rather than in a new satellite series. While new and improved Landsat satellites benefited the users, they did not substitute for an operational system because they were not accompanied by a commitment to ongoing availability of standard data.

While Department of the Interior leaders could push NASA to plan an operational satellite, the two agencies had little success persuading the Office of Management and Budget to provide funding because Landsat had not yet developed the constituency of users required by the budget office. A NASA representative promised in testimony to Congress in 1973 that Landsat would become financially self-sufficient, but that day appeared to be a long way away.[13] In fact, the attendees at a 1973 Department of the Interior meeting discussed the "apparent lack of demand for the ERTS [Landsat] type data sufficient to justify an operational system."[14] Lack of broad user support was sometimes painfully visible; in 1976 the Department of Agriculture responded to questions from a Senate committee with the statement that the Department was not convinced that a Landsat-type system was "cost-effective and [could] satisfy USDA data needs."[15] The need to justify an operational earth resources satellite system by stimulating wider use of Landsat data put additional pressure on NASA's technology transfer program.[16]

Gradual Transition to a More Operational Program

A number of problems discouraged potential users from making increasing operational use of Landsat data. Paradoxically, obstacles arose both because the project was too experimental and because it was too operational. Some users avoided investing in expensive systems and time-consuming training when they were uncertain that Landsat data would continue to be available. Others had doubts over whether Landsat data was adequate for particular applications and argued that they needed an improved satellite. Yet the commitment of a significant number of users to Landsat data had created a small but important constituency that would be offended if the data they used ceased to be available.

Users, particularly those who would need continuing, repetitive coverage, were reluctant to invest in systems to use Landsat data when there was no assurance that an operational earth resources satellite system would follow. As the National Research Council's Committee on Remote Sensing Programs for Earth Resources

Surveys explained, "Until data flow continuity is assured, the principal users probably will continue to be experimenters and those whose interest is largely confined to static phenomena."[17] This posed only minor problems for relatively simple data analysis techniques similar to those used for aerial photographs, which required relatively small investments. However, technologically advanced methods for using the data, in which NASA scientists and engineers saw so much promise, faced a barrier of uncertain availability of data as well as the barrier of user dislike of advanced technology.

On the other hand, some customers were already using the data operationally, which constrained the flexibility NASA managers wanted in an experimental project. The earth resources program had to "maintain a reasonable continuity of data for which the user community has proven needs and capabilities"[18] in order not to alienate the small constituency that had already formed. When NASA made plans to test new sensors on *Landsat 4*, it had to argue for them with the Department of the Interior:

> At this stage in the Landsat series, NASA should not want to nor are we justified in going forward with a proposal for another satellite and payload essentially identical to that of the first three missions. But, in essence, that is what DOI proposes that we do. We believe, just as DOI does, that there should be as much continuity of data as practical, and we agree that there should be maximum utilization of the existing data coming from the present systems, but not at the expense of sacrificing the extension of current technology state-of-the-art as expeditiously as possible.[19]

NASA managers wanted to conduct a continuing research program aimed at developing a better satellite, while the Department of the Interior wanted an operational satellite as soon as possible.

Disagreements caused particularly severe problems when it came time to design the next Landsat satellite, *Landsat 4*. Because the Office of Management and Budget would not approve funds to build separate experimental and operational systems, the agencies involved had to compromise both among different data needs and between an operational and an experimental system.

The contradictory preferences of the various agencies became particularly clear when the Office of Management and Budget proposed eliminating the multispectral scanner from *Landsat 4*. Budget agency officials argued that the scanner was ready for operational use and therefore did not belong on an experimental satellite. A memo written by a representative from the National Oceanic and Atmospheric Administration describing an interagency meeting in March 1977 observed:

> The major concern is the wording in OMB's allocation letter to NASA which states: "continued flight of the multispectral scanner instrument is contingent upon NASA's ability to obtain funding from Landsat data users. As the Landsat program evolves, it is expected that increased emphasis will be given to obtaining cost recovery from domestic and foreign users." The user agencies are afraid that any pitch they make on carrying the MSS [multispectral scanner] will be construed by OMB as a commitment to assist in the funding.[20]

Not only were the agencies reluctant to set a precedent by supporting part of an experimental satellite, they also feared that the end result would be a budget cut. If the agencies identified money from their approved program that they were willing to spend on the scanner, the budget bureau "might just take the money away without providing it to NASA."[21] The agencies finally changed their strategy. They argued that the scanner was required not as an operational instrument but rather as an essential research instrument to provide standard data for comparison with the output of the thematic mapper. On this basis the Office of Management and Budget finally gave NASA the necessary approval to include the scanner on *Landsat 4* but not on the backup satellite.[22]

The disagreements over the technology for *Landsat 4* were hard to resolve because that satellite represented a step toward an operational system. Compromises that had been made for the experimental system were rescrutinized for an operational system. The first few Landsats had been NASA's responsibility, but the user agencies would be expected to make good, practical use of an operational system. Therefore, each agency again pushed for changes in the technology that would make the system more useful for their specific purposes.

The user agencies opposed some changes in the Landsat system and urged others. At an August 1977 meeting, the Department of the Interior representative complained about the new orbit for *Landsat 4*, which made it retrievable by the Shuttle orbiter but meant that even the scanner data would be difficult to compare with that from previous missions. The same person objected that the thematic mapper did not suit the needs of the Department of the Interior (it better served the need of the Department of

Agriculture). Leonard Jaffe, NASA Deputy Associate Administrator for Applications, tried to blunt the controversy by explaining that the thematic mapper "was designed to define the parameters for a future operational sensor and . . . was not necessarily or even probably the ultimate operational sensor."[23] Interior wanted a different type of operational sensor, a solid-state device called a *pushbroom scanner*, similar to the technology later proposed for the French and Japanese earth resources satellites. The pushbroom scanner used a simpler design than the thematic mapper and therefore would probably be less expensive and more reliable. But it did not provide the infrared data required by the Department of Agriculture or the fine resolution desired by a number of users.[24]

Control of the Operational System

Discussions of *Landsat 4* as a proto-operational system also gave new urgency to questions of whether the government would make a commitment to an operational earth resources satellite system and, if so, what organization would manage that system. In the end the pressure exerted by interested agencies and the small group of other users forced a commitment to an operational Landsat system. However, instead of choosing a manager for the system from among the competing agencies, the White House decided that the system should be turned over to private industry. At that point this analysis must end; the privatization of Landsat has been a tortuous process and it is too soon to provide a historical evaluation.

Even in 1977 and 1978 the interested agencies had come to no consensus over how best to manage an operational earth resources satellite system. The EROS program of the Department of the Interior put in an early bid for control but did not receive support from the Secretary of the Interior.[25] Meanwhile, pressure for an operational system continued to mount, bringing the issue to the attention of the White House. In an attempt to resolve the question, the Office of Science and Technology Policy conducted a major study of policy options for civilian remote-sensing satellites. In the final report of that study (June 1978), three consultants—Bruno Augenstein, Willis H. Shapley, and Eugene B. Skolnikoff—concluded that the president should make a commitment to an operational earth resources satellite system and suggested that NASA have responsibility for that system.[26]

User agency leaders opposed NASA control of the operational system, at least without organizational mechanisms to ensure that NASA would be responsive to user needs. The Departments of Agriculture and the Interior clearly had to be involved, and even state users expressed concern. A task force of state representatives reported the following in 1980:

For a number of years, the States have argued for a formal role in national satellite remote sensing policy. If a remote sensing system is to be responsive to State needs, the States must be involved in the management of the system. The states are a special class of users; they share responsibility for the management of the nation's national resources with the federal government. Many of the State's needs for remote sensing data result from Federal mandates and directives.[27]

Questions of management of the operational system involved not just who would control the system but also who would pay for it. The states wanted to obtain earth resources data as a service provided by the federal government and funded by federal appropriations rather than having to pay the full cost themselves in the form of higher prices for data.

In the end the White House resolved the debate over control over an operational earth resources system by first giving the system to a neutral agency and then transferring it to private industry. In November 1979 President Carter issued Presidential Directive 54, which gave the National Oceanic and Atmospheric Administration temporary responsibility for operational earth satellites.[28] NOAA already administered the successful operational weather satellite program, so giving NOAA control of operational earth resources satellites not only avoided the controversy among NASA, the Department of the Interior, and the Department of Agriculture but also took advantage of existing expertise. NOAA assumed responsibility for *Landsat 4* after its initial testing and started planning for an eventual fully operational system.

However, while most observers assumed that weather satellites would remain a government program, President Carter initiated a radically different approach for earth resources satellites. The Carter administration called for NOAA to develop a plan for turning earth resources satellites over to private industry.[29] The predicted large benefit-cost ratio for Landsat implied that the data could be sold at a high enough price to cover the cost of the entire satellite system. It thus fell within Carter's plan to remove from government control functions that private industry could handle. This involved a choice between different visions of the program: An operational earth resources satellite system could be run either as a profit-making enterprise or as a government service.

A significant number of users argued vehemently that earth resources satellites should remain a government service. A 1980 report by a task force of state representatives stressed that Landsat data should remain inexpensive: The government should not attempt to recover the costs of research or even the capital cost of the operational system. The authors explained that "the establishment and operation of a land remote sensing satellite system is viewed as a public service in the same context as census, cartographic, geological and meteorological data which are provided by the Federal government."[30] Better management of parkland, prevention of floods by more accurate prediction of spring runoff, and many other benefits did not bring immediate financial return.[31] In addition, control by private industry, even under government regulation, would raise questions about how the company running the earth resources satellite program might use its access to the data to take advantage of developing countries. Studies by the Comptroller General and the Congressional Office of Technology Assessment argued that private ownership would result in an overall reduction of the benefits earth resources satellites provided to the nation and the world.[32]

Despite opposition by many government agencies and industry concerns about whether an operational earth resources system would make a profit, President Ronald Reagan signed the Land Remote Sensing Commercialization Act of 1984 on July 17, 1984. The law required data to be sold on a nondiscriminatory basis, provided a mechanism for temporary government subsidy and continuing government research and development, and required the Department of Commerce to maintain a data archive.[33] However, it proved difficult to find companies interested in taking on earth resources satellites, and transfer was delayed by disagreements over the need for government subsidy. In October 1985 the Earth Observing Satellite Corporation (Eosat, a joint venture of Hughes Aircraft Co. and RCA Astro-Electronics) took over the Landsat program, but some observers still doubted whether commercialization would be successful.[34]

In 1979 it appeared that NASA's earth resources satellite program was ready to make the transition from an experimental to an operational system. Nine years later, the future of the project was still uncertain, because the fundamental disagreements that had plagued the program from the beginning had still not been fully solved. Instead, the situation was further complicated by a new political mood critical of big government, which resulted in pressure to turn the whole program over to private industry. As in the early stages of the project, the negotiation between the technology developer and the potential users to define the new system took place in an environment shaped in critical ways by the interests of other players within the government system.

Epilogue

The future of operational earth resources satellite systems growing out of Landsat depends not only on finding an effective integration of the interests of the government and private enterprise and of research and development teams and users but also on outside factors such as the development of an infrastructure for space commercialization. This study of the experimental phases of Landsat cannot predict eventual success or failure, but it can help us understand the role of the users and other interest groups in the development, testing, and diffusion of new technology.

Landsat provides a good example of how the pattern of competing interests between technological developers and users changes in each phase of the life cycle of a technology. During the research phase NASA wanted support but not interference from the user agencies. The Department of the Interior quickly became interested in Landsat and tried to turn the project toward what scientists at that department saw as the quickest, simplest, useful system, but NASA resisted. Interior's attempt to take over earth resources satellites by creating the EROS program resulted more in hostility than in progress toward consensus between developers and users. Despite the publicity generated by early controversies between NASA and user agencies, user involvement during development was relatively limited. People with actual operational responsibility—the resource managers in federal, state, and regional agencies—had no role in the early stages of the project.

During the later stages of development and design of the satellite the debate among groups interested in Landsat became focused on selection of sensors and the question of whether the satellite project would be primarily experimental or operational. Landsat necessarily had both experimental aspects as an investigation of how a new system would work and operational aspects because it provided data intended for routine use by resource managers. However, the definition of Landsat as operational or experimental was the focus of interagency tensions. NASA defined an experimental project as one that could develop technology for its own sake rather than always being responsive to the requirements of the users. The Bureau of the Budget required that the project be experimental so that there would be no commitment to another government program until the idea had proved successful. The Department of the Interior, and other enthusiasts, fought for as operational a project as possible to increase their control and the influence of the users.

The balance between developers and users changed during testing and development of new applications, but many problems continued. After a slow start, NASA leaders realized that the agency would have to support applications research because successful applications were essential to building a constituency that would support the evolution of an experiment into an ongoing program. The resulting effort to support applications research was not always successful, in part because NASA's style did not lend itself to working with the appropriate users. NASA managers had experience in supporting scientific research, but they had little sense of the operational needs of resources managers. Rather, as in the Large Area Crop Inventory Experiment, NASA projects tended to develop more sophisticated technology than the users wanted. Those who might best have developed applications often did not trust the space agency. These prospective users were suspicious because they had not been involved from the beginning in the project, because NASA leaders seemed condescending, and because they assumed that any technology developed in the space program was too complex for routine, economical use.

User agency suspicions proved to be even more of a problem when the time came to diffuse new applications of Landsat data to large numbers of potential users. At first NASA failed to recognize that transfer of new techniques required a program to actively convince individual users, not just good publicity. Initially, the Department of the Interior's EROS Data Center had the best results in encouraging use, because that department had a history of earning the trust of resources managers. Yet NASA could not hope for much trust from the users; state government users recognized particularly clearly that the project would not be useful to them unless they had some control over it. The natural solution might have been an operational system controlled by the users, but Executive Office interests prevented that. The question of whether Landsat had proved successful delayed any sort of operational system, and that in turn discouraged those potential users who did not want to make an investment to use satellite data that might be discontinued. Different interest groups had different opinions about whether Landsat was a success or a failure depending on each group's definition of what the system should do for it. The operational system was also delayed by the controversy over whether earth resources satellites should be run as a business or be subsidized by the government as a public service.

Bringing a new technology into use is by necessity a slower process than NASA had hoped for Landsat, but the space agency also made mistakes that discouraged potential users. The data from the satellite quickly proved valuable to foreign countries, the extractive industry, and some government agencies, such as the Foreign Agricultural Service and the Corps of Engineers. The value of the data was perhaps best demonstrated by the French and Japanese decisions to launch more advanced earth resources satellites. Yet, the lack of operational use by agencies like the Department of Agriculture's Statistical Reporting Service and state and regional resource management agencies has been disappointing to NASA policymakers. If Landsat had been a success in the eyes of all interested parties, it would probably have led to an operational earth resources satellite before 1980. Instead, it is still not certain whether Landsat will lead to a significant ongoing program. As of 1989, the ongoing program is tenuous, and could easily fade away if the company that has taken over Landsat is unable to compete with the successful French SPOT satellite and other potential competitors that will be launched in the next few years.

This study suggests that historians of technology need to consider more carefully a number of issues concerning the role of users in defining new technology. We cannot look at the question only in terms of whether a technology will sell. Nor can the role of the users be simplified entirely to the argument that any new system will be more successful if its eventual users are given a voice and a stake in the system by early involvement in its design.

These approaches are certainly valid, but this study suggests that different understandings of the technology are at stake in addition to user commitment and consideration of user needs. The conflict over when Landsat should be declared operational provided an example of a fight directly over definitions, which reflected a whole range of user interests. Even more important, the repeated lack of effective communication between NASA and the Department of the Interior and other users usually reflected not purposeful noncooperation but rather different definitions of the proposed technological system. Different groups had very different visions of what pattern the process of development would take, of what kinds of data would be most useful, of what types of methods would be developed to use the satellite data, and of how an eventual operational system should be designed and managed.[1]

Only further study can show whether the Landsat case is typical or unusual. While Landsat was not clearly a failed technology, it was at least much more difficult to develop and bring into use than expected. Its story seems to suggest that incommensurable expectations of a technology result in winners and losers or in unsatisfactory compromises.[2] Historians need to look at similar issues in the cases of more successful technological change to understand how different interest groups can better communicate.

Notes

Chapter 1

1. For an introduction to some of these approaches, see Wiebe E. Bijker, Thomas P. Hughes, and Trevor Pinch, eds., *The Social Construction of Technological Systems: New Directions in the Sociology and History of Technology* (Cambridge: MIT Press, 1987); Donald MacKenzie and Judy Wajcman, eds., *The Social Shaping of Technology: How the Refrigerator Got Its Hum* (Milton Keynes, Philadelphia: Open University Press, 1985); and Bruno Latour, *Science in Action* (Cambridge: Harvard Univ. Press, 1987).

2. Exceptions may be found in Bruno Latour, *Science in Action*, and in Wiebe E. Bijker and Trevor J. Pinch's work on the social construction of the bicycle in Trevor J. Pinch and Wiebe E. Bijker, "The Social Construction of Facts and Artifacts: Or How the Sociology of Science and the Sociology of Technology Might Benefit Each Other," in *The Social Construction of Technological Systems*, especially pp. 28–39.

3. Bruno Latour defines interests as "what lie *in between* actors and their goals, thus creating a tension that will make actors select only what, in their own eyes, helps them reach these goals amongst many possibilities." *Science in Action*, pp. 108–109.

4. Latour, *Science in Action*, p. 110. However, the space agency has rarely succeeded in keeping its interest groups in line to the degree that Latour treats as an essential part of strategy for innovation. It may be that interagency politics is such an effective mechanism for negotiating between interest groups that one group is rarely able to dominate.

5. Abbot Payson Usher, *A History of Mechanical Invention* (New York: McGraw-Hill, 1954). Jacob Schmookler, *Invention and Economic Growth* (Cambridge: Harvard Univ. Press, 1966). S. C. Gilfillan, *The Sociology of Invention* (Cambridge: MIT Press, 1963, first published 1935). See especially T. P. Hughes, "Inventors: The Problems They Choose, The Ideas They Have, and the Inventions They Make," in Patrick Kelly and Melvin Kranzberg, eds., *Technological Innovation* (San Francisco: San Francisco Press, 1978).

6. Thomas P. Hughes, *Networks of Power: Electrification in Western Society, 1880–1930* (Baltimore: Johns Hopkins Univ. Press, 1983).

7. See Thomas P. Hughes, "The Development Phase of Technological Change," *Technology and Culture* 17 (1976): 423–431.

8. See, for example, Edwin Mansfield, *The Economics of Technological Change* (New York: W. W. Norton, 1968). Some historians call the production-and-sale stage of

technological change *innovation*, but this terminology can be confusing because others use innovation with a broader meaning.

9. An extremely valuable introduction to this literature is provided by Everett M. Rogers, *Diffusion of Innovations*, 3d ed. (New York: The Free Press, 1983).

10. For an interesting attempt to tease apart these different components see Edward W. Constant II, "The Social Locus of Technological Practice: Community, System, or Organization," in Bijker, Hughes, and Pinch, *The Social Construction of Technological Systems*.

11. The debate over second-generation weather satellites provides a good example of this tension (see chapter 2). It is important to note that in some cases entrepreneurs do successfully create or manage the need for their innovation. Donald MacKenzie provides a very suggestive analysis of a case of this kind in "Missile Accuracy: A Case Study in the Social Processes of Technological Change," in Bijker, Hughes, and Pinch, *The Social Construction of Technological Systems*.

12. A more detailed discussion of this point can be found in Leonard R. Sayles and Margaret K. Chandler, *Managing Large Systems* (New York: Harper and Row, 1971).

13. In addition to Bijker, Hughes, and Pinch, *The Social Construction of Technological Systems*, see Edward W. Constant, II, *The Origins of the Turbojet Revolution* (Baltimore: Johns Hopkins Univ. Press, 1980), which adapts for technology the theory developed in Thomas Kuhn, *The Structure of Scientific Innovation*, 2d ed. (Chicago: Univ. of Chicago Press, 1970). For other efforts in the same direction see Rachel Lauden, ed., *The Nature of Technological Knowledge: Are Models of Scientific Change Relevant?* (Dordrecht: Reidel, 1984).

14. Latour, *Science in Action*, makes this point particularly clearly.

15. For the social construction of science see for example: Barry Barnes and David Edge, eds., *Science in Context: Readings in the Sociology of Science* (Cambridge: MIT Press, 1982) and Karen Knorr-Cetina and Michael J. Mulkay, eds., *Science Observed: Perspectives on the Social Studies of Science* (Beverley Hills: Sage, 1983).

16. See Bijker, Hughes, and Pinch, *The Social Construction of Technological Systems*.

17. John M. Logsdon, *The Decision to Go to the Moon* (Cambridge: MIT Press, 1970). See also W. Henry Lambright, *Presidential Management of Science and Technology: The Johnson Presidency* (Austin: Univ. of Texas Press, 1985).

18. One example of this type of study at its best is R. Cargill Hall, *Lunar Impact* (Washington, D.C.: NASA SP-4210, 1977). See also Lloyd S. Swenson, Jr., J. Grimwood, and C. Alexander, *This New Ocean: A History of Project Mercury* (Washington, D.C.: NASA SP-4201, 1966; Barton C. Hacker and J. Grimwood, *On the Shoulders of Titans: A History of Project Gemini* (Washington, D.C.: NASA SP-4203, 1977; Edward C. Ezell and L. Ezell, *On Mars: Exploration of the Red Planet 1958–1978* (Washington, D.C.: NASA SP-4212, 1984); Roger E. Bilstein, *Stages to Saturn: A Technical History of the Apollo/Saturn Launch Vehicles*, (Washington, D.C.: NASA SP-4206, 1980.

19. The classic example is weather forecasts, which are subsidized by the government because they provide widespread benefits. Those benefits are diffuse in the sense that they are difficult to measure, and it would be impossible to fund the Weather Bureau by charging fees to the individuals who benefit from better forecasts. More formally, the concept of public good has many different defini-

tions, as illustrated in Glendon Schubert, *The Public Interest: A Critique of the Theory of a Political Concept* (Glencoe, Ill.: The Free Press, 1960). My working definition, which I think follows the most common usage at NASA, is that the public good is the long-range benefit of the society, balancing the interests of organized and unorganized interest groups, as opposed to the short-range benefit of industry or other individual interest groups. (Professor Schubert would probably characterize this definition as somewhere between social engineering and psychological realism.)

20. Attitudes have also changed over time. Following a tradition established in the nineteenth century, the federal government has tended to fund technology only as an expense of accomplishing certain clear government missions, not as an investment for the nation. See A. Hunter Dupree, *Science in the Federal Government: A History of Policies and Activities to 1950* (Cambridge: Harvard Univ. Press, 1957).

21. W. Henry Lambright, *Governing Science and Technology* (New York: Oxford Univ. Press, 1976).

22. A strongly stated and useful discussion of the effect of agency self-interest on Landsat can be found in W. Henry Lambright, "ERTS: Notes on a 'Leisurely' Technology," *Public Science* (Aug.-Sept. 1973): 1–8.

23. Dupree, *Science in the Federal Government.*

Chapter 2

1. U.S. Senate, Committee on Commerce, Science, and Transportation, *National Aeronautics and Space Act of 1958, As Amended, and Related Legislation,* 95th Congress, 2d session (Washington, D.C.: GPO, 1958), p. 1. For an excellent analysis of Eisenhower's attitude toward the space program see Walter A. McDougall, . . . *the Heavens and the Earth: A Political History of the Space Age* (New York: Basic Books, 1985).

2. For a discussion of this process see McDougall, . . . *the Heavens and the Earth.*

3. Homer E. Newell, *Beyond the Atmosphere: Early Years of Space Science* (Washington, D.C.: NASA SP-4211, 1981), chapter 12.

4. For a recent example of the continuing argument see James A. Van Allen, "Space Science, Space Technology, and the Space Station," *Scientific American* 254/1 (January 1986): 32–39.

5. For examples of the Pentagon point of view see John B. Medaris, *Countdown for Decision* (Princeton, N.J.: Van Nostrand, 1959) and Hans Mark, *The Space Station: A Personal Journey* (Durham, N.C.: Duke Univ. Press, 1987).

6. John M. Logsdon, *The Decision to Go to the Moon: Project Apollo and the National Interest* (Cambridge: MIT Press, 1970).

7. Signs of declining support could be seen as early as 1964, when NASA leaders began to be asked to justify their budgets and goals, but it was only in 1969 that they began to scale back their plans for the future in major ways.

8. Alex Roland, *Model Research: The National Advisory Committee for Aeronautics, 1915–1958* (Washington, D.C.: NASA SP- 4103, 1985).

9. See Arnold S. Levine, *Managing NASA in the Apollo Era* (Washington, D.C.: NASA SP-4102, 1982).

10. This tendency was greatest in the case of the Jet Propulsion Laboratory, well described in Clayton R. Koppes, *JPL and the American Space Program* (New Haven, Conn.: Yale Univ. Press, 1982). See also Elizabeth A. Muenger, *Searching the Horizon: A History of Ames Research Center 1940–1976* (Washington, D.C.: NASA SP-4304, 1985).

11. Newell, *Beyond the Atmosphere*, pp. 250–252. The conflict between NASA headquarters and a somewhat anomalous center is a major theme of Koppes, *JPL and the American Space Program*.

12. Logsdon, *The Decision to Go to the Moon*.

13. Newell, *Beyond the Atmosphere*.

14. W. Henry Lambright, *Governing Science and Technology* (New York: Oxford Univ. Press, 1976).

15. My information on weather satellites comes from Richard LeRoy Chapman's excellent dissertation: "A Case Study of the U.S. Weather Satellite Program: The Interaction of Science and Politics" (Syracuse University, 1967).

16. This improvement has been less dramatic than many expected, not because of the quality or quantity of image data provided by weather satellites, but because of the lack of a model of the atmosphere exact enough to make reliable weather predictions even from plentiful data.

17. My discussion of communications satellites is based mostly on Jonathan F. Galloway, *The Politics and Technology of Satellite Communications* (Lexington, Mass.: Lexington Books of D.C. Heath and Co., 1972), and Delbert D. Smith, *Communication Via Satellite: A Vision in Retrospect* (Leyden, Boston: A. W. Sijthoff, 1976). I also found helpful Michael E. Kinsley, *Outer Space and Inner Sanctums: Government, Business, and Satellite Communications* (New York: John Wiley, 1976), a Nader report; J. R. Pierce, *The Beginnings of Satellite Communications* (San Francisco: San Francisco Press, 1968), giving the AT&T view; and Roger A. Kvam, "Comsat: The Inevitable Anomaly," in Stanford A. Lakoff, ed., *Knowledge and Power: Essays on Science and Government* (New York: The Free Press, 1966).

18. Levine, *Managing NASA in the Apollo Era*, p. 216.

19. Technically, Landsat produces images, not photographs (a photograph is defined as an image formed directly on a recording medium by electromagnetic radiation from the subject, thus excluding even television cameras). *Picture* is used here as a nontechnical synonym for *image*. Many technical people object to calling Landsat images pictures because picture may be seen as a synonym for photograph (there is a particular concern to differentiate Landsat data from the photographs taken by reconnaissance satellites). It is important to remember that Landsat data could be prepared for study not only in the form of images reproduced on photographic film but also in the form of computer tape.

Chapter 3

1. Grover Heiman, *Aerial Photography: The Story of Aerial Mapping and Reconnaissance*, Air Force Academy Series (New York: Macmillan, 1972), pp. 30–31.

2. I. B. Holley, *Ideas and Weapons: Exploitation of the Aerial Weapons by the United States During World War I; A Study in the Relationship of Technological Advance, Military Doctrine, and the Development of Weapons* (Hamden, Conn.: Archon Books, 1971).

3. Heiman, *Aerial Photography*, pp. 58–59, 84.

4. "Aerial Mapping and Unemployed Engineers," *Science-Supplement* n.s. 78 (Dec. 29, 1933): 7–8. Marshall S. Wright, "The Aerial Photographic and Photogrammetric Activities of the Federal Government," *Photogrammetric Engineering* 5 (1939): 168–176.

5. Robert S. Quackenbush, Jr., "Development of Photo Interpretation," in *Manual of Photographic Interpretation* (Virginia: American Society of Photogrammetry, 1960), p. 7.

6. Constance Babington-Smith, *Air Spy: The Story of Photo Intelligence in World War II* (New York: Harpers, 1957).

7. Heiman, *Aerial Photography*, pp. 77–79, 122–128.

8. Anthony Kenden, "U.S. Reconnaissance Satellite Programmes," *Spaceflight* (July 1968): 259.

9. Heiman, *Aerial Photography*, pp. 122–128, 132–143.

10. RAND was a project initiated by the U.S. Army Air Force by contract with the Douglas Aircraft Company in March 1946. It became an independent nonprofit corporation in 1948. The fullest discussion available so far in unclassified form of the role of RAND in early satellite plans is Merton E. Davies and William R. Harris, *RAND's Role in the Evolution of Balloon and Satellite Observation Systems and Related U.S. Space Technology*, RAND Publication Series R-3692-RC (Santa Monica, Calif.: The RAND Corporation, Sept. 1988). For a general discussion of the origins of satellite plans see R. Cargill Hall, "Early U.S. Satellite Proposals," *Technology and Culture* 4 (1963): 410–434.

11. Kenden, "U.S. Reconnaissance Satellite Programmes," pp. 243–244.

12. Walter A. McDougall, *. . .the Heavens and the Earth: A Political History of the Space Age* (New York: Basic Books, 1985), part 2.

13. For an example of these attitudes see John B. Medaris with Arthur Gordon, *Countdown for Decision* (New York: Putnam, 1960).

14. William E. Burrows, *Deep Black: Space Espionage and National Security* (New York: Random House, 1986), pp. 91, 104.

15. Ibid. p. 110. Burrows quotes a source who claims that the system could provide a resolution on the order of twelve inches (p. 111).

16. Kenden, "U.S. Reconnaissance Satellite Programmes," pp. 246–250.

17. Ibid., p. 247.

18. Office of the Director, Goddard Space Flight Center, to J. Naugle and L. Jaffe, "Additional Comments on the Possible Use of Lunar Orbiter Residual Hardware for an Earth Resources Technology Satellite," July 5, 1968; Thomas Ragland's files, Goddard Space Flight Center, Greenbelt, Md.

19. The air force sponsored an early study of television cameras for reconnaissance satellites, but the study concluded in 1957 that the system was not worth development at the time. Burrows, *Deep Black*, p. 90.

20. Burrows, *Deep Black*, pp. 232–233.

21. Ibid., p. 213–214.

22. Kenden, "U.S. Reconnaissance Satellite Programmes," p. 252. Burrows, *Deep Black*, pp. 227–229, 244–251.

23. Burrows, *Deep Black*, pp. 111, 215, 234, 235, 238.

24. In fact, resolution has a number of different technical definitions. NASA usually defined resolution by the most closely spaced black and white parallel bars that could be distinguished in a test image.

25. Peter C. Badgley to Deputy Director, Space Applications Programs and Director of Meteorology, "Guidelines for Meeting with DoD Relative to Geography Program," Mar. 30, 1966; "Documentation—Earth Resources" folder, History Office, NASA Hq.

26. Interviews with J. Robert Porter, EarthSat Corporation, Bethesda, Md., July 28, 1978, and with Leonard Jaffe, NASA Hq, Washington, D.C., July 27, 1978.

27. Interview with Fabian Polcyn, Environmental Research Institute of Michigan, Ann Arbor, Mich., May 11, 1981.

28. Interviews with George J. Zissis, Symposium on Remote Sensing of Environment, Ann Arbor, Mich., May 13, 1981 and with Marvin Holter, Environmental Research Institute of Michigan, Ann Arbor, Mich., May 15, 1981.

29. *Proceedings of the First Symposium on Remote Sensing of Environment, 13, 14, 15 Feb. 1962* (Ann Arbor, Mich.: Infrared Laboratory, Institute of Science and Technology, Univ. of Michigan, 1962).

30. *Proceedings of the Second Symposium on Remote Sensing of Environment, 15, 16, 17, Oct. 1962* (Ann Arbor, Mich.: Infrared Laboratory, Institute of Science and Technology, Univ. of Michigan, 1962).

31. Ibid., p. 4.

32. Ibid., p. 427.

33. Ibid., p. 428.

34. This provides an interesting contrast with radio astronomers in the period immediately after World War II, who sucessfully transfered military technology into a new field of science. See David O. Edge and Michael J. Mulkay, *Astronomy Transformed: The Emergence of Radio Astronomy in Britain* (New York: John Wiley, 1976). Part of the difference lies in greater military resistance, but a more entrenched scientific community may also have been less willing to take risks to promote a new field.

35. Telephone conversation with Robert Piland, Johnson Space Center, Houston, March 9, 1981.

36. J. S. Butz, Jr., "Man's Eye in the Sky is Sharp Indeed," *Air Force and Space Digest* 52 (Jan. 1969): 56–58.

37. Peter C. Badgley to the Associate Administrator, "'Visual Acuity' in Space," July 15, 1965; biography file for Peter C. Badgley, History Office, NASA Hq.

38. Leonard Jaffe to Congressional Liaison Director, "Telephonic Request for Information re Apollo and Gemini Photography," Dec. 19, 1968, "Documentation—Earth Resources" folder, History Office, NASA Hq.

39. Paul D. Lowman, Jr., and Herbert A. Tiedemann, *Terrain Photography From Gemini Spacecraft: Final Geological Report* (Greenbelt Md.: Goddard Space Flight Center #X-644-71-15, Jan. 1971), p. iii.

40. Ibid., p. 66.

41. James O. Spriggs to the record, "Imagery From Space," Feb. 16, 1967; "*Landsat 1* Documentation" folder, History Office, NASA Hq.

42. S. Fred Singer and Robert W. Popham, "Non-Meteorological Observations From Weather Satellites," *Astronautics and Aerospace Engineering* 1 (April 1963): 89–92.

43. Archibald Park and Oscar Weinstein, "A Proposal for a Multispectral TV Camera System Experiment for Earth Resources Application," [Spring 1968]; no folder, 076-184 (2), RG 255, Washington National Records Center.

Chapter 4

1. Hall shows that lunar science won a voice in NASA's space science program as early as 1959, but NASA hired a significant number of geologists only when the agency started gearing up for Apollo. R. Cargill Hall, *Lunar Impact: A History of Project Ranger* (Washington, D.C.: NASA SP-4210, 1977) pp. 6–15.

2. This was a ticklish subject because Apollo was widely perceived as a scientific project, but in fact science took a back seat to national prestige objectives and the safety of the crew. This often frustrated the scientific community, whose public criticism could be very damaging to NASA. For more discussion of this issue see Homer E. Newell, *Beyond the Atmosphere: Early Years of Space Science* (Washington, D.C.: National Aeronautics and Space Administration, 1980), pp. 208–209.

3. Interview with Peter C. Badgley, Office of Naval Research, Arlington, Va., Aug. 7, 1978.

4. Peter C. Badgley to BFA-2, Procurement Requests dated Dec. 28, 1964, Mar. 16, 1965, July 20, 1965, Aug. 11, 1965, Nov. 5, 1965, and Nov. 23, 1965; biography file for Peter C. Badgley, History Office, NASA Hq.

5. "Earth-Oriented and Lunar-Oriented Experiments for Manned Orbital Missions," [Fall 1974]; biography file for Peter C. Badgley, History Office, NASA Hq.

6. For example: Robert W. Pease, "Plant Tissue and the Color Infrared Record," Status Report 3, Technical Report 2, U.S. Department of Interior Contract 14-08-0001-10674, Leonard W. Bowden, Principal Investigator, no date; History Office, Johnson Space Center, Houston, Texas.

7. Badgley, Procurement Request, July 20, 1965.

8. Badgley, Procurement Request, Mar. 29, 1965.

9. Badgley, Procurement Request, Nov. 5, 1965.

10. Badgley, Procurement Request, Aug. 17, 1965. Radar sensors are an example of a technology with clear potential that was not selected for use on Landsat. Most simply, radar could be used to determine the distance between an aircraft or spacecraft and the ground, and thus map the contours of the terrain. It also provided information from another region of the spectrum that might differentiate things that looked the same in visible light, and could penetrate clouds, vegetation and sometimes soil. For a good summary of the capabilities of radar sensors and the programs that have used them see Harry Joyce, "Space Radar for Remote Sensing" *Spaceflight* 28 (March 1986): 134–137. Radar was not included in Landsat

because of the decision to build a small satellite oriented toward the known needs of the users; at the time sensor decisions were made, radar had not been proved valuable to a strong enough constituency of users.

11. Interview with Fabian Polcyn, ERIM, Ann Arbor, Mich., May 11, 1981.

12. K. Staenz, F. J. Ahern, and R. H. Brown, "The Influence of Illumination and Viewing Geometry on the Reflectance Factor of Agricultural Targets," paper presented at the Fifteenth International Symposium on Remote Sensing of Environment, May 11–15, 1981, Ann Arbor, Mich.

13. Alden P. Colvocoresses to the record, "Technical Matters Relative to Departure of Colonel Colvocoresses," Aug. 4, 1967; Alden Colvocoresses's files, U.S. Geological Survey.

14. R. N. Colwell, "Some Practical Applications of Multiband Spectral Reconnaissance," *American Scientist* 49 (March 1961): 9–36.

15. In the view of satellite proponents this advantage disappeared when a satisfactory sensor had been developed and tested; at that point flying it in space in a satellite that would last for years could be a less expensive way of gathering large amounts of data than an aircraft program.

16. Bernard T. Nolan, "NASA's Airborne Research Program," April 1978; Bernard Nolan's files, NASA Hq.

17. "Earth Resources Aircraft Program, Manned Spacecraft Center, Sept. 19, 1968" [transcript of a press conference]; History Office, Johnson.

18. Peter C. Badgley, "Current Status of NASA's Natural Resources Program," *Proceedings of the Fourth Symposium on Remote Sensing of Environment*, April 12–14, 1966 (Ann Arbor, Mich.: Infrared Physics Laboratory of Willow Run Laboratories, Institute of Science and Technology, Dec. 1966). Interview with Leo Childs, Johnson, Sept. 10, 1979.

19. Willis B. Forster to the Deputy Associate Administrator (attn. Col. J. C. Damon), "Initiation of Request to U.S. Navy for Loan of P3A Aircraft to NASA," Sept. 8, 1965; "SM Reading File 1965" folder, box 52, accession 74-663, record group 255, Washington National Records Center. Apparently, the Johnson Space Center was offered a U-2 but refused to accept it because of lingering hostility over the claim made initially by the White House that the U-2 piloted by Gary Powers shot down over the Soviet Union was on a NASA research mission.

20. Interview with Robert B. MacDonald, Johnson, Sept. 10, 1979.

21. Interview with Allen H. Watkins, EROS Data Center, Sioux Falls, S.Dak., May 8, 1981.

22. Ibid.

23. For an interesting discussion of the development of the U-2 see William E. Burrows, *Deep Black: Space Espionage and National Security* (New York: Random House, 1986).

24. Some of this program was carried out in the Apollo and Skylab programs, and perhaps Space Station will prove its value, but so far, earth resources experiments from spacecraft that carry people have produced results less satisfactory to most resource managers than data from automated satellites. Most important, Apollo and Skylab experiments did not last long enough to provide the systematic

coverage that would have made the data widely useful, although they did provide early space tests of some advanced sensors. In particular, a system of cameras flown on *Apollo 9* in early 1969 provided images similar to those Landsat would produce, thus preparing scientists to interpret the Landsat images. Robert N. Colwell, et al., *Monitoring Earth Resources From Aircraft and Spacecraft* (Washington, D.C.: NASA SP-275, 1971).

25. Some people at NASA realized the potential political value of Landsat at this time, but the momentum of the agency kept them from being heard. A 1967 draft memo from the office of the Assistant Administrator for Policy stated: "ERS [Earth Resources Survey] offers NASA a unique and timely opportunity to furnish the president and Congress with demonstrable evidence that the space program can be effectively applied to critical national and international problems such as poverty, overpopulation, urban stabilization and the enhancement of natural resources. If it is to survive in today's budgetary environment, NASA must be responsive to these very real problems, as well as continue with its vital efforts to reach out and explore the solar system." Draft Memorandum to Mr. Webb, Dr. Seamans, Dr. Newell, "Issues Re: The Earth Resources Survey Program," attachment to Jacob E. Smart to distribution, "Earth Resources Survey Program," Oct. 3, 1967; "Earth Resources Documentation," History Office, NASA Hq.

26. W. David Compton and Charles D. Benson, *Living and Working in Space: A History of Skylab* (Washington, D.C.: NASA SP-4208, 1983), particularly pp. 182–194.

27. Paul Dickson, *Out of This World: American Space Photography* (New York: Delacorte Press, 1977), p. 36. W. Henry Lambright, "ERTS: Notes on a 'Leisurely' Technology," *Public Science* (Aug.–Sept. 1973): 4–5.

28. Edgar M. Cortright, "Space Science and Applications: Where We Stand Today," *Astronautics and Aeronautics* 4 (June 1966): 42–48.

29. For a discussion of technological momentum see Thomas P. Hughes, "Technological Momentum in History: Hydrogenation in Germany 1900–1933," *Past and Present* no. 44 (1969): 106–132.

30. Badgley, "Current Status of NASA's Natural Resources Program."

31. Exactly what they were referring to is not clear. Many scientific experiments were cut from Apollo for safety's sake after the fatal fire in an Apollo capsule in early 1967. When *Apollo 9* was launched in 1969, it included an earth photography experiment but not a scanner. Cancellation of scientific experiments planned for Apollo and cutbacks of funding also caused dramatic changes in plans for Apollo Applications. In early 1967 NASA scheduled a special earth resources mission, *Apollo Application 1A*, to be launched in late 1968, but that plan was canceled in August 1967. Compton and Benson, *Living and Working in Space*, pp. 83–87.

32. "Prepared by Jaffe and Badgley at Seamans's Request: NASA Natural Resources Program," May 13, 1966; "Documentation—Earth Resources" folder, History Office, NASA Hq.

33. Homer E. Newell, *Beyond the Atmosphere: Early Years of Space Science* (Washington, D.C.: NASA SP-4211, 1980).

34. Peter C. Badgley to the record, "Relative Funding Support by NASA and the Earth Resources User Agencies Over the Next Several Years," Aug. 10, 1966; "Documentation—Earth Resources" folder, History Office, NASA Hq.

35. Ibid.

36. Peter C. Badgley to the record, "Interior Department News Release on Earth Resources Observation Satellite (EROS)," Sept. 29, 1966; "Documentation—Earth Resources" folder, History Office, NASA Hq.

Chapter 5

1. Homer E. Newell to Dr. Thomas B. Nolan, Jan. 19, 1975; "SM Reading file" folder, box 52, accession number 74-663, record group 255, Washington National Records Center.

2. Interview with Charles J. Robinove, EROS Program Office, Reston, Va., July 31, 1978. "EROS Program—Issue Paper," attachment to David S. Black to James E. Webb, Nov. 21, 1967; "Related Sciences 3, ERTS/EROS, NASA/Interior Collaboration," 77- 0677 (33), RG 255, WNRC.

3. Interview with Willis H. Shapley, American Association for the Advancement of Science, Washington, D.C., Aug. 13, 1979.

4. Edward Z. Gray to Director, Manned Space Science, "Pete Badgley's White Paper on Remote Sensing," Oct. 18, 1965; Peter C. Badgley to Deputy Director, Space Applications Programs and Director of Meteorology, "Guidelines for Meeting with DoD Relative to Geography Program," Mar. 30, 1966; "Response to Supplementary Questions Raised by Associate Deputy Administrator in Connection with Schultze (BoB) Letter to Webb Re FY '68 Budget Backup Materials," Mar. 30, 1966; all in "Documentation—Earth Resources" folder, History Office, NASA Hq.

5. Leonard Jaffe to the record, "Commentary Delivered by Mr. Leonard Jaffe at the Airlie House Planning Seminar, June 1966," July 8, 1966; biography file on Leonard Jaffe, History Office, NASA Hq. Peter C. Badgley to the record, "Relative Funding Support by NASA and the Earth Resources User Agencies Over the Next Several Years," Aug. 10, 1966; "Documentation—Earth Resources" folder, History Office, NASA Hq.

6. Badgley, "Relative Funding Support," Aug. 10, 1966.

7. Ibid.

8. "Space Applications Program, May 23, 1967," program review document; Donald P. Rogers's files, NASA Hq.

9. Robert G. Reeves to the record, "Meeting at the U.S. Geological Survey (USGS), 10 A.M., Aug. 25, 1966," Aug. 31, 1966; "Documentation—Earth Resources" folder, History Office, NASA Hq.

10. Leonard Jaffe to the Deputy Administrator, "Meeting at the U.S. Geological Survey, August 25, 1966, Regarding Remote Sensing and South America," Sept. 6, 1966; "Documentation—Earth Resources" folder, History Office, NASA Hq.

11. Reeves, "Meeting at the U.S. Geological Survey," Aug. 31, 1966.

12. Robinove interview. Interview with William Fischer, EROS Program Office, Reston, Va., Aug. 8, 1978.

13. Robinove interview.

14. Fischer interview.

15. Office of the Secretary, U.S. Department of the Interior, News Release, "Earth's Resources to be Studies from Space," Sept. 21, 1966; "EROS Program—Creation" folder, EROS Program Office files.

16. Fischer interview.

17. Ibid.

18. Ibid.

19. Shapley interview.

20. Charles P. Boyle to John F. Clark, "Highlights of NASA Hearing Before the House Subcommittee on Space Science and Applications, March 19, 1969," Mar. 20, 1969; "ERS, ERTS" folder, Information Processing Division Files, Goddard Space Flight Center, Greenbelt, Md.

21. Interview with Marvin Holter, ERIM, Ann Arbor, Mich., May 15, 1981.

22. Peter C. Badgley to the record, "Interior Department News Release on Earth Resources Observation Satellite (EROS)," Sept. 29, 1966; "Documentation—Earth Resources" folder, History Office, NASA Hq.

23. Robert C. Seamans, Jr., to Stewart L. Udall, Sept. 22, 1966; "Related Sciences 3, ERTS/EROS, NASA/Interior Collaboration" folder, History Office, NASA Hq.

24. Ibid.

25. Charles F. Luce to Robert C. Seamans, Jr., Oct. 7, 1966; "Related Sciences 3, ERTS/EROS, NASA/Interior Collaboration" folder, 77-0677 (33), RG 255, WNRC.

26. Charles F. Luce to Robert C. Seamans, Jr., Oct. 21, 1966, with attachment "Operational Requirements for Global Resource Surveys by Earth-Orbital Satellites: EROS Program;" "Related Sciences 3, ERTS/EROS, NASA/Interior Collaboration" folder, 77-0677 (33), RG 255, WNRC.

27. Leonard Jaffe to the record, "Meeting at Organization of American States (OAS) on Applicability of Remote Sensor Satellite Programs for Latin America, November 3, 1966," Nov. 15, 1966; "*Landsat 1* Documentation" folder, History Office, NASA Hq.

28. Robert C. Seamans, Jr., to Charles F. Luce, April 7, 1967; "*Landsat 1* Documentation" folder, History Office, NASA Hq.

29. "Specific Comments on October 21 Letter From C. F. Luce to R. C. Seamans," Jan 14, 1967, attachment to Robert C. Seamans, Jr., to Charles F. Luce, April 7, 1967; "*Landsat 1* Documentation" folder, History Office, NASA Hq.

30. John Rhea, "Earth Resources Satellite Far From Reality," *Technology Week* 20 (Feb. 13, 1967): 34–37.

31. L. Jaffe to Dr. Newell, "Article on the Earth Resources Satellite in *Aerospace Technology*, October 9 Issue," Oct. 13, 1967. "Related Sciences 3, ERTS/EROS, NASA/Interior Collaboration" folder, 77-0677(33), RG 255, WNRC.

32. W. T. Pecora to the Under Secretary, "Status of EROS Program," June 15, 1967; William Fischer's Significant Documents file, EROS Program Office.

33. Charles F. Luce to Robert C. Seamans, Jr., April 24, 1967; "Related Sciences 3, ERTS/EROS, NASA/Interior Collaboration" folder, 277-0677(33), RG 255, WNRC. Robert C. Seamans to Charles F. Luce, June 1, 1967; William Fischer's Significant Documents file, EROS Program Office.

34. Pecora, "Status of EROS Program."

35. Stewart L. Udall to James E. Webb, Sept. 9, 1967; James E. Webb to Stewart L. Udall, Dec. 1, 1967; both in "*Landsat 1* Documentation" folder, History Office, NASA Hq.

Chapter 6

1. T. M. Ragland to D. C. Clatterbuck, "Information Requested by Mr. Frank Hammil, Jr., Counsel for House Committee on Science and Aeronautics," Dec. 2, 1968; document #002735, box 3, accession number 81-416, record group 255, Washington National Records Center.

2. For the history of Ranger (and a discussion of the failures that plagued the program) see R. Cargill Hall, *Lunar Impact: A History of Project Ranger* (Washington, D.C.: NASA SP- 4210, 1977).

3. "RCA Bidding for ERS Job with Vidicon," *Aerospace Technology* 21 (Oct. 23, 1967): 43.

4. Edward Risley to Staff Officers of Divisions, Offices, and Committees, National Research Council, "Display of Camera/Reproducer Systems for Spacecraft Applications," Nov. 14, 1967; #002563, 81-416(2), RG 255, WNRC.

5. "RCA Reports Camera Breakthrough for EROS Applications," *Space Daily* (Oct. 13, 1967): 228. "RCA Bidding for EROS Job with Vidicon," p. 42.

6. R. A. Stampfl to the record, "Frequency Allocation for Operational Earth Resources Service Use," Nov. 22, 1967; #002576, 81-416(2), RG 255, WNRC.

7. Oscar Weinstein to Harvey Ostrow, "Trip to RCA-AED on 11/8 and 11/9, 1967, Contract NAS 5-10339," Nov. 20, 1967; #02562, 81-416(2), RG 255, WNRC.

8. T. A. George to Morris Tepper, "Request for Additional FY 1968 Funding on Work Unit 160-75-02-06-51, TV Multispectral Camera," April 5, 1968; "Documentation—Earth Resources" folder, History Office, NASA Hq.

9. Rune Evaldson, "Willow Run Laboratories: Institute of Science and Technology, the University of Michigan," March 1970; I. J. Sattinger's files, Environmental Research Institute of Michigan, Ann Arbor, Mich.

10. Interview with Fabian Polcyn, ERIM, Ann Arbor, Mich., May 11, 1981.

11. Polcyn interview. Interview with Marvin R. Holter, ERIM, Ann Arbor, Mich., May 15, 1981. This forced declassification was ironic because earth resources satellites might have been much less successful without the availability of multispectral scanner technology for civilian use. In addition, the lack of interest in the earlier patent showed the importance of the research done by the military in exploring the potential of remote sensing.

12. "P. C. 713-42695: Justification for Noncompetitive Procurement," June 6, 1967; #03284, 81-416(4), RG 255, WNRC.

13. Archibald B. Park and Oscar Weinstein, "A Proposal for a Multispectral TV Camera System Experiment for Earth Resources Application," no date but in a box of proposals dated Jan.-Mar. 1968; no folder, 76-184(2), RG 255, WNRC.

14. Interview with Archibald Park, Washington, D.C., Aug. 3, 1978. In 1969 NASA engineers were so nervous about the complexity of the Hughes scanner that they at one point decided to fly both it and a scanner with much more limited potential developed by Hycom because they "were not sure if either one of the two scanners will work out." William Nordberg to Wilfred E. Scull, "Multi-spectral Scanners for

ERTS, Mar. 26, 1969; #002935, 81-416 (3), RG 255, WNRC. Nordberg, ERTS Study Scientist at Goddard, was arguing for a greater commitment to the more promising multispectral scanner.

15. William Nordberg to M. Schneebaum, "Multichannel Infrared, Scanning Radiometer," Oct. 3, 1968; #002662, 81-416(1), RG 255, WNRC.

16. J. R. Porter, "Summary Minutes: Earth Resources Survey Program Review Committee, 28 Feb. 1969;" "Documentation—Earth Resources" folder, History Office, NASA Hq.

17. "Development of Operational Satellite Systems," attachment to John R. Borchert to William T. Pecora, Aug. 25, 1967; "*Landsat 1* Documentation" folder, History Office, NASA Hq.

18. Jacob E. Smart to Dr. Seamans, "Meeting with Representatives of Departments of Agriculture and Interior Re Earth Resources," Sept. 11, 1967; "Related Sciences 3, ERTS/EROS, NASA/Interior Collaboration" folder, 77-0677(33), RG 255, WNRC.

19. G. Suits to R. Evaldson, "Suggested Security Guide for Surveillance and Target Acquisition Technology," Oct. 27, 1967; G. Suit's files, ERIM, Ann Arbor, Mich.

20. "NASA, DoD Locked in Heated Debate on Earth Surveillance, *Aerospace Daily*, Mar. 13, 1968.

21. T. A. George to Leonard Jaffe, "DoD December 21, 1967, Memorandum on Airborne IR Imagery Security Classification," April 1, 1968; "Documentation—Earth Resources" folder, History Office, NASA Hq. Texas Instruments, "Nomogram for Determining Classification of Infrared Systems," no date; #002346, 81-416(1), RG 255, WNRC.

22. D. Williamson, "H. Brown Telecon Outline," Feb. 2, 1977; "Fletcher Correspondence" folder, History Office, NASA Hq. Cooperation was essential because the military space program was at least as large as the civilian program.

23. C. P. Cabell to J. Robert Porter, Jr., "Remote Sensors for Earth Oriented Purposes," May 9, 1968; "Documentation—Earth Resources" folder, History Office, NASA Hq.

24. W. Nordberg to members, GSFC Earth Resources Working Group, "Discussion of Follow-on Missions to *ERTS A* & *B*," Mar. 16, 1970; "*Landsat 1*" folder, Information Processing Division files, Goddard Space Flight Center, Greenbelt, Md.

25. Interview with Allen H. Watkins, EROS Data Center, Sioux Falls, SD, May 8, 1981.

26. Interview with Alden P. Colvocoresses, U.S. Geological Survey, Reston, Va,, Aug. 7, 1979.

27. Archibald B. Park to Leonard Jaffe, Sept. 11, 1968; "ERSPRC, 1968, Dr. Naugle" folder, David Williamson's files, NASA Hq.

28. "EROS Applications Resolution Requirements," attachment to William A. Fischer to Robert J. Porter [sic], Sept. 12, 1968; "ERSPRC, 1968, Dr. Naugle" folder, David Williamson's files, NASA Hq.

29. Fischer to Porter, Sept. 12, 1968.

30. Robert B. McEwen to Alden Colvocoresses, "October 1969 Revision to NASA Document S-701-P-3, 'Design Study Specification for the Earth Resources Tech-

nology Satellite *ERTS A* and *B*,'" Jan 21, 1970; "*Landsat 1* Documentation" folder, History Office, NASA Hq.

31. W. A. Hovis to N. W. Spencer, "Earth Resources Scanner Parameters," June 6, 1969; #03198, 81-416(4), RG 255, WNRC.

32. Ibid.

33. William Nordberg to multiple addressees, Oct. 6, 1970; #005324, 81-416(10), RG 255, WNRC.

34. W. A. Hovis to Tom Ragland, "ERTS Scanner and RBV Wavelength Interval Selection," July 28, 1969; #003426, 81- 416(5), RG 255, WNRC.

35. "Field Support Program for ERTS," draft, July 1, 1969; #03341, 81-416(5), RG 255, WNRC.

36. Roger M. Hoffer to D. A. Landgrebe, "Wavelength Band Selection," Nov. 1, 1973; History Office, Johnson Space Center, Houston, Tex.

37. James C. Fletcher to Frank G. Zarb, 27 Sept. 1974; "OMB Budget Justification" folder, H. Mannheimer's files, NASA Hq.

38. Nicholas M. Short to Dr. C. C. Kraft, "High Resolution Pointable Imager (HRPI) Review," Mar. 25, 1975; History Office, Johnson.

39. For an interesting case showing many of the same patterns in an even more politicized setting see Robert W. Smith and Joseph N. Tatarewicz, "Replacing a Technology: The Large Space Telescope and CCDs," *Proceedings of the IEEE* 73 (1985): 1221–1235.

Chapter 7

1. While Landsat met special hostility, it was not unusual for NASA projects in the late 1960s to suffer so severely at the hands of the Bureau of the Budget. The cutbacks of the Apollo Applications program during the same period are well documented in W. David Compton and Charles D. Benson, *Living and Working in Space: A History of Skylab* (Washington, D.C.: NASA SP 4208, 1983). The role of the Office of Mangement and Budget in the Space Shuttle decision a few years later was equally dramatic. See John M. Logsdon, "The Space Shuttle Program: A Policy Failure?" *Science* 232 (May 30, 1986): 1099–1105.

2. The policy-setting role of the Bureau of the Budget and, as it was later known, the Office of Management and Budget during the Nixon administration deserve extensive examination, but that is not the focus here.

3. During the late 1960 and early 1970s the fiscal year ran from July 1 to June 30, starting before the calendar year—fiscal 1969 ran from July 1968 to June 1969.

4. Interview of William Lilly by John Mauer, NASA Hq., Washington, D.C., Oct. 20, 1983; Johnson Space Center Historians Source files.

5. Homer Newell, *Beyond the Atmosphere* (Washington, D.C.: NASA SP-4211), p. 325.

6. "Earth Resources Survey Briefing: Outline," Sept. 21 [1967]; F. W. Godsey biography file, History Office, NASA Hq. "Response to Supplementary Questions Raised by Associate Deputy Administrator in Connection with Shultze (BoB) Letter to Webb Re FY'68 Budget Backup Materials," draft, Mar. 30, 1966; "Documentation—Earth Resources" folder, History Office, NASA Hq.

7. Jacob E. Smart to Dr. Seamans, "Earth Resources Sensing Capability," Sept. 5, 1967; "Related Sciences 3, ERTS/EROS, NASA/Interior Collaboration" folder, box 33, accession number 77- 0677, record group 255, Washington National Records Center.

8. George M. Low to Assistant Administrator for Public Affairs, "Project Names," Nov. 8, 1974; Fletcher correspondence, History Office, NASA Hq. Helen T. Wells, Susan H. Whitely, and Carrie E. Karegeannes, *Origins of NASA Names* (Washington, D.C.: NASA SP-4402, 1976), p. 43.

9. "EROS Program Issue Paper," attachment to David S. Black to James E. Webb, Nov. 21, 1967; "Related Sciences 3, ERTS/EROS, NASA/Interior Collaboration" folder, 77-0677 (33), RG 255, WNRC.

10. Economics, Science, and Technology Division (D. E. Crabill) to Director, Bureau of the Budget, "Appeals Session for NASA," Dec. 18, 1967; "BoB 1969," box 59, 60.18a, RG 51, National Archives. The other cuts appealed were in the Apollo, Large Solid Rocket, and Aircraft Technology programs.

11. Handwritten notes in the Bureau of the Budget files titled "Agency Head Mtg. 12-19-67" describe three points made by Newell: "Using 'forcing function' argument. Agrees to try to prove NASA wrong on this argument if we hold ERTS. Webb wants to talk about it in Congress whether we approve it or not." The last point is marked with a star; since the NASA administrator was a member of the executive branch and a presidential appointee, he or she was expected to maintain a united front in support of the budget that the president sent to Congress. "Agency Head Mtg. 12-19-67;" "BoB 1967," box 59, 60.18a, RG 51, National Archives. The forcing function argument was explained in a letter from Webb to Budget Director Schultze: "From past experience, only real data, taken under actual environmental conditions, can be the catalyst and forcing function needed to establish the research baseline that can allow a proposed operation system to be properly assessed." James E. Webb to Charles L. Schultze, Dec. 15, 1967; "BoB 1969," box 59, 60.18a, RG 51, National Archives.

12. Interview with John M. DeNoyer, U.S. Geological Survey, Reston, Va., Aug. 7, 1969.

13. Joseph E. Karth, "Earth Resources Surveys—An Outlook on the Future," presented at an IEEE meeting, Feb. 13, 1969, attachment to Bernard P. Miller to Tom Ragland, March 3, 1969; #002824, 81-416 (3), RG 255, WNRC.

14. "Seeking a Better View," *Business Week* (July 13, 1968).

15. Thomas M. Ragland to Wilfred E. Scull, "68-2 POP for ERTS Study Funds," July 25, 1968; Thomas Ragland's files, NASA Hq.

16. John M. DeNoyer, "Summary of the History, Role, and Activities of the Earth Resources Observation Systems (EROS) Program: Contributions to a Congressional Study of U.S. Civilian Space Activities," July 1978; William Hemphill's files, EROS Program Office.

17. RES (Richard Speier) to the Director, Bureau of the Budget, "NASA Preview Materials," June 12, 1968; "P1-2, box 248, 61.1a, RG 51, National Archives.

18. "EST Division Staff Paper on Earth Resources Satellites," Nov. 18, 1968; "P1-4, box 248, 61.1a, RG 51, National Archives.

19. Charles P. Boyle to John F. Clark, "Highlights of NASA Hearing Before the House Subcommittee on Space Science and Applications, Mar. 19, 1969," Mar. 20, 1969; "ERS, ERTS" folder, Information Processing Division, Goddard Space Flight Center, Greenbelt, Md.

20. Ibid.

21. Natural Resources Programs Division (Monte Canfield) to the Director, Bureau of the Budget, "Earth Resources Satellite (EROS)," Dec. 12, 1968; "P1-4," box 248, 61.1a, RG 51, National Archives.

22. DeNoyer, "Summary of the History."

23. "Interior Department Appeal of 1971 Budget Allowance: EROS," received at NASA November 25, 1969; "Related Sciences 3, ERTS/EROS, NASA/Interior Collaboration" folder, 77-0677 (33), RG 255, WNRC.

24. Ibid.

25. "NASA Fy 1971 Budget Reclama of BoB Tentative Allowance," [late November 1969]; "*Landsat 1* Documentation" folder, History Office, NASA Hq.

26. DeNoyer, "Summary of the History."

27. For example, Arlen J. Large, "How Backers of a Technology Satellite Induced Ford to Overrule His Advisors, Provide Money," *Wall Street Journal*, Feb. 20, 1975.

28. George M. Low, acting NASA Administrator, to Donald B. Rice, Assistant Director, Office of Management and Budget, April 22, 1971; "ERTS/ERS Documents/Fletcher 71-77" folder, History Office, NASA Hq.

29. See Arnold S. Levine, *Managing NASA in the Apollo Era* (Washington, D.C.: NASA SP-4102, 1982), p. 189.

30. Suggested by Homer Newell in conversation.

31. Interview with William A. Fischer, EROS Program Office, Reston, Va., Aug. 8, 1978.

32. Amrom H. Katz, "Reflections on Satellites for Earth Resources Surveys: Personal Contributions to a Summer Study," RAND Paper P-3753, 1967. "Comments on Aircraft and Space Survey Systems," Appendix E of *Useful Applications of Earth Oriented Satellites: Sensors and Data Systems* (Washington, D.C.: National Academy of Sciences, 1969).

33. Amrom H. Katz, "Let Aircraft Make Earth-Resource Surveys," *Astronautics and Aeronautics* 7 (June 1969): 60–68. B. F., "ERS Resolution: Old Numbers Out," *Astronautics and Aeronautics*, 8 (April 1970): 13–14. Katz, "ERS Resolution: Old Numbers in Again?" *Astronautics and Aeronautics* 8 (Aug. 1970): 53–55.

34. Amrom H. Katz, "A Retrospective on Earth Resource Surveys: Arguments About Technology, Analysis, Politics, and Bureaucracy," *Photogrammetric Engineering and Remote Sensing* 42 (Feb. 1976): 189–99.

35. Katz probably had more than one motivation; one theme that runs through his writings is is a series of complaints about using expensive technology when a simpler solution was available. He had earlier played the role of skeptic about classified reconnaissance satellites. Katz, then working at the Reconnaissance Laboratory at Wright Field, was a member of a group who set out in 1951 "to prove that this proposed project [a RAND proposal for a reconnaissance satellite with a television camera] was ridiculous." After some difficulty in finding a lens poor

enough "to simulate the performance of this rotten TV sensor cruising at 300 miles altitude" the team produced results that were more promising than Katz had expected. Not long thereafter Katz joined RAND to work on reconnaissance. Quotes from Katz in Merton E. Davies and William R. Harris, *RAND's Role in the Evolution of Balloon and Satellite Observation Systems and Related U.S. Space Technology*, (Santa Monica, Calif.: RAND Corporation, September 1988), pp. 27–31.

36. Fischer interview.

37. RES (Richard Speier) to the Director, Bureau of the Budget, "NASA Preview Materials," June 12, 1968; "P1-2" folder, box 248, 61.1a, RG 51, National Archives.

38. "NASA FY 1971 Budget Reclama."

39. Ibid.

40. "Outline for ERTS Presentation: 2 December 1969"; "*Landsat 1* Documentation" folder, History Office, NASA Hq.

41. "NASA Comments on Memorandum for Mr. William Anders by Dr. David D. Elliott, Subject: ERTS Resolution," draft, Apr. 3, 1971; "Related Sciences 3, ER, ERS, ERTS, 1971-1972" folder, 77-0677 (33), RG 255, WNRC.

42. More research needs to be done on the origins of Bureau of the Budget use of cost-benefit arguments to challenge the space program. By late 1969 OMB was criticizing NASA's space shuttle plans on cost-benefit grounds, despite Department of Defense support for that program.

43. "EROS Program Issue Paper."

44. "NASA: Economic Applications in Space; Analysis of Changes in Program Requirements," BoB 1969 budget packet; box 59, 60.18a, RG 51, National Archives.

45. Interview with Charles J. Robinove, EROS Program Office, Reston, Va., July 31, 1978.

46. Robert P. Mayo to Walter J. Hickel, Apr. 14, 1970; William Fischer's Significant Documents File, EROS Program Office.

47. James C. Fletcher, handwritten note on GML[ow] to Jim F[letcher] Oct. 29, 1974; "ERTS/ERS Documents/Fletcher 71-77" folder, History Office, NASA Hq.

48. Earth Satellite Corporation and Booz-Allen Applied Research, "Evaluation of Economic and Social Benefits and Costs of Data from *ERTS-A*," July 1972. Environmental Research Institute of Michigan, "Evaluation of Economic, Environmental, and Social Cost and Benefits of Future Earth Resources Survey Satellite Systems," Nov. 1972.

49. Karth, "Earth Resources Surveys."

50. Mayo to Hickel, Apr. 14, 1970.

Chapter 8

1. Interview with J. Robert Porter, Jr., Earth Satellite Corp., Bethesda, Md., July 16, 1978.

2. Wilfred E. Scull to Special Programs Office, "Semi-Annual Report of Improved Manpower Management," May 26, 1969; document #003110, box 4, accession number 81-416, record group 255, Washington National Records Center.

3. David Williamson, Jr., "Summary Minutes: Interagency Coordinating Committee: Earth Resources Survey Program, May 11, 1972;" "Minutes of the ICCERSP" folder, vol. 1, David Williamson's files, NASA Hq. Christopher C. Kraft, Jr., to Dale D. Myers, June 15, 1972; History Office, Johnson Space Center, Houston, Tex.

4. Interview with Clifford E. Charlesworth, Johnson, Mar. 18, 1981.

5. William Nordberg to T. George, "Coordination with User Agencies—ERTS," [May 1969]; "ERTS Users Working Group" folder, Information Processing Division files, Goddard Space Flight Center, Greenbelt, Md.

6. Ibid.

7. For example, NASA engineers and policy-makers made decisions about storing data on board the satellite with the users in mind but with a minimum of actual consultation. Landsat satellites were intended to take images of the entire world, but few of the facilities of NASA's worldwide net of radio reception stations had the capacity to receive large amounts of data at the high speed required for Landsat. Therefore, NASA decided to include a tape recorder on the satellite. Landsat would transmit data as they were collected over the United States, but it would store data obtained from the rest of the world and play them back from tape to a receiving station in Alaska. There were some doubts about this plan's feasibility, because tape recorders had proved to be unreliable on a number of scientific satellites. It was not clear that a tape recorder could be built that would be adequate for so large an amount of rapid data storage and Landsat's projected long life (J. Hayes to W. Scull, "ERTS—Video Tape Recorder," Oct. 15, 1969; no control number, 81-416(8), RG 255, WNRC). In the fall of 1969 NASA investigated saving money by eliminating the tape recorder, but Goddard and NASA Headquarters made the decision to keep it in the plan so that users would have more complete world coverage (Werner Gruhl to Robert E. Bourdeau, "Video Tape Recorder vs. Existing Land Stations: ERTS Extra U.S. Imagery Collection," Dec. 24, 1969; #003945, 81-416(6), RG 255, WNRC).

8. Alden P. Colvocoresses to Research Coordinator, EROS program, "Liaison with Goddard on *ERTS-A* and *-B*," 23 Jan. 1970; "*Landsat 1* Documentation" folder, History Office, NASA Hq.

9. John E. Naugle to the administrator, "Interagency Coordination of the Earth Resources Survey Program," June 25, 1971; packet of material from D. Rogers, History Office, NASA Hq.

10. James C. Fletcher to Donald B. Rice, July 1, 1971; "Fletcher Correspondence" folder, History Office, NASA Hq.

11. H. R. Syverson to A. H. Wessels, "Request for an Informational/Planning Purposes Proposal for an Earth Resources Satellite," May 12, 1967; #02280, 81-416(1), RG 255, WNRC.

12. J. Goett to Daniel G. Mazur, Dec. 3, 1967; #002098, 81-416(1), RG 255, WNRC.

13. Interview with William A. Fischer, EROS Program Office, Reston, Va., Aug. 8, 1978.

14. John E. Naugle to Assistant Administrator for Legislative Affairs, "Answers to Frank Hammill's Questions Relating to Earth Resources Survey (ERS) Staff Study," Dec. 9, 1968; "Documentation—Earth Resources" folder, History Office, NASA Hq.

230   Notes to pages 102–110

15. William Nordberg to Thomas Ragland, "ERTS Procurement," Feb. 19, 1969; #03167, 81-416(4), RG 255, WNRC. Interview with Stanley R. Addess, Goddard, Aug. 2, 1979.

16. See chapter 9 for a more detailed discussion of the data processing system.

17. Addess interview.

18. Interview with John M. DeNoyer, U.S. Geological Survey, Reston, Va., Aug. 7, 1979. Interviews with Stanley C. Freden and Thomas M. Ragland, Goddard, Aug. 2, 1979.

19. Clare F. Farley to the record, "Selection of Contractor for Award of *ERTS A & B* (Phase D) Procurement," Sept. 5, 1970; "Related Sciences 3, ERTS/Contract Award" folder, 77-0677(33), RG 255, WNRC.

20. Ibid.

21. E. J. Bosley, GE, to W. E. Scull, "Bi-Weekly Status report," Nov. 23, 1971; no folder, 77-267(2), RG 255, WNRC.

Chapter 9

1. For another example of a situation in which NASA chose a technologically conservative course see James E. Tomayko, "NASA's Manned Spaceflight Computers," *Annals for the History of Computing* 7 (January 1985): 7–18.

2. For striking examples of how NASA had to be conservative in its technological choices see James E. Tomayko, *Computers in Spaceflight: The NASA Experience*, Encyclopedia of Computer Science and Technology, vol. 18, supplement 3 (New York: Marcel Dekker, 1987. For an example of the institutional dangers of failure see R. Cargill Hall, *Lunar Impact: A History of Project Ranger. History Series* (Washington, D.C.: National Aeronautics and Space Administration, 1977).

3. Alden P. Colvocoresses to the record, "Technical Matters Relative to Departure of Colonel Colvocoresses," Aug. 4, 1967; A. P. Colvocoresses's files, U. S. Geological Survey, Reston, Va.

4. Robert Stephens to Curtis M. Stout, "ERTS Data Processing," Goddard Space Flight Center Record of NASA Headquarters Telephone Contact, Nov. 8, 1968; "ERTS NDPF thru 1972" folder, Information Processing Division, Goddard Space Flight Center, Greenbelt, Md. Leonard Jaffe to Robert R. Gilruth, Wilmot Hess, and Robert O. Piland, "ERS Data Handling," Apr. 8, 1969; "Documentation—Earth Resources" folder, History Office, NASA Hq. "NASA Position Paper on the Earth Resources Survey Program," Nov. 1970; "Related Sciences 3, ER, ERS, ERTS 1971-72" folder, box 33, accession number 77-0677, record group 255, Washington National Records Center.

5. William A. Anders to George M. Low, enclosing a memo from David Elliott to Low on "ERTS Resolution," Feb. 24, 1971; "Related Sciences 3, Remote Sensing, ERTS/ERS/EROS/ERTS, Jan.-Mar. #1, 1971" folder, 77-0677 (33), RG 255, WNRC.

6. Thomas M. Ragland to the record, "*ERTS-A & -B* Data Acquisition, Processing, Analysis, & Archival Services," Dec. 30, 1968; "ERTS NDPF thru 1972" folder, Information Processing Division, Goddard. Robert L. Bell to John J. Gorman, with enclosed draft of sections 5 and 6 of an OST study report on data handling, May

11, 1970; "ERTS Users Working Group" folder, Information Processing Division, Goddard.

7. See chapter 2.

8. Interview with John Y. Sos, Goddard, July 25, 1979.

9. "Consolidated Budget for the National Earth Resources Survey Program—FY 1972," attachment to John E. Naugle to Donald B. Rice, Dec. 8, 1970; "Earth Resources Documentation" folder, History Office, NASA Hq.

10. Interview with John M. DeNoyer, Geological Survey, Aug. 7, 1979.

11. Alden P. Colvocoresses, "Report of Meeting on Data Handling," Feb. 10, 1969; and Raymond W. Fary, Jr., to Attendees, Research Coordinator, and Chairman of Working Groups, "ERTS Data Requirements Conference February 10, 1969," Feb. 12, 1969; both in "ERTS Users Working Group" folder, Information Processing Division, Goddard.

12. William Nordberg to Wilfred E. Scull, "Review of User Agency Requirements for *ERTS A* in Response to 17 September Meeting at GSFC," Oct. 14, 1970; "ERTS NDPF thru 1972" folder, Information Processing Division, Goddard.

13. William Nordberg to William Fischer, Oct. 19, 1970; "ERTS Users Working Group" folder, Information Processing Division, Goddard.

14. Thomas M. Ragland to J. W. Clifton, Apr. 5, 1972; and Thomas M. Ragland to Clifford A. Spohn, Apr. 5, 1972; both in "NDPF Workload" folder, 77-395(1), RG 255, WNRC.

15. William Nordberg to multiple addressees, "Minutes of ERTS User Agency Meeting—13 March 1970," Apr. 14, 1970; "ERTS Users Working Group" folder, Information Processing Division, Goddard. Goddard Space Flight Center, "Design Study Specifications for the Earth Resources Technology Satellite: *ERTS-A* and *-B*," Revision B, S-701-P-3, Jan. 1970; 77-395(1), RG 255, WNRC. Ad Hoc Group on ERS Data Management, "Earth Resources Technology Satellite (ERTS) Data Management," Feb. 1970; 77-395(2), RG 255, WNRC.

16. W. T. Pecora to John E. Naugle, May 19, 1970; and Clifford M. Hardin to T. O. Paine, May 15, 1970; both in "ERTS NDPF thru 1972" folder, Information Processing Division, Goddard. Interview with Stanley C. Freden, Goddard, Aug. 2, 1979.

17. Homer Newell to Donald B. Rice, July 22, 1971; Donald B. Rice to Homer E. Newell, Aug. 9, 1971; and "Statement of NASA Earth Resources Survey Imagery Policy," enclosure to Homer Newell to Donald B. Rice, Mar. 10, 1972; all in "Related Sciences 3, Photography, 1970-" folder, 75-637(3), RG 255, WNRC.

18. R.W. Gutmann to James C. Fletcher Jan. 27, 1975; History Office, "Fletcher Correspondence, NASA.

19. *Landsat Data Users Handbook* (Greenbelt, Md.: NASA-Goddard 76-SDS-4258, Sept. 2, 1976).

20. "Earth Resources Technology Satellite, NASA Data Processing Facility, Interface Control Document for NDPF/Users Interface: Image Products and Catalogs," ICD 6108, distribution date Oct. 7, 1971; #004523, 77-269(1), RG 255, WNRC.

21. Robert L. Bell to members, ERS Data Management Group, "Corrections and Additions to Minutes of May 6 Meeting," May 28, 1970; #004523, 81-416(7), RG 255, WNRC.

22. William C. Webb to Frank A. Keipert, "Improvements in Timeliness for Landsat Image Processing," June 16, 1975. "ERS, ERTS" folder, Information Processing Division, Goddard. General Electric, "*ERTS-1* Systems Performance Assessment Report: The First 90 Days," 77-395(1), RG 255, WNRC. "Type I Progress Report: *ERTS-A*, Predict Ephemeral & Perennial Range Quantity and Quality during Normal Grazing Season, *ERTS-A* Proposal No. SR 147," Jan. 1973; "User Complaints" folder, 77-395(1), RG 255, WNRC. George H. Ludwig to John F. Clark, "ERTS NDPF Visit by C. Mathews," Feb. 17, 1972; "ERTS NDPF thru 1972" folder, Information Processing Division, Goddard.

23. Sos interview.

24. John Y. Sos to Luis Gonzales, "NDPF Image Generation Augmentation," April 16, 1973; "ERTS NDPF" folder, Information Processing Division, Goddard.

25. "User Bulletin—Dark Negatives," Dec. 1, 1972; "ERTS NDPF thru 1972" folder, Information Processing Division, Goddard. T. F. Mackin to D. Holmes, "User Feedback," Nov. 6, 1972; "User Complaints" folder, 77-395(1), RG 255, WNRC. Alexander F. H. Goetz to William Nordberg, Nov. 13, 1972, "ERTS NDPF thru 1972" folder, Information Processing Division, Goddard. Anthony E. Salerno to John Sos, Nov. 20, 1972, "User Complaints" folder, 77- 395(1), RG 255, WNRC. John Y. Sos to William Nordberg, "New NDPF Film Product for Investigators with Test Sites in Barren Regions," Jan 4, 1973; and John Y. Sos to multiple addressees, "Addition of a New NDPF Film Product," Jan 22, 1973; both in "New Film Product" folder, 77-395(1), RG 255, WNRC.

26. That does not prove that the decision was wrong. The technology changed so fast that the system may have been too risky in 1970 and the only way to go in 1975.

27. John Y. Sos to J. Naugle and J. DeNoyer, "Evaluation of Proposals for All-Digital Precision Pressing Techniques Development," Aug. 30, 1971; "ERS & ERTS Proposal Evaluation" folder, Information Processing Division, Goddard. George H. Ludwig to Mr. Truszynski, "Response to Question from Dr. Dineen," Feb. 7, 1972; "ERTS Headquarters" folder, Information Processing Division, Goddard. Sos interview. Frank A. Keipert to Office of the Director, "Digital Image Processing Contracts, Apr. 27, 1973; "*Landsat 1*" folder, Information Processing Division, Goddard.

28. Ralph Bernstein, "Scene Correction (Precision Processing) of ERTS Sensor Data Using Digital Image Processing Techniques," paper presented at the third ERTS Symposium, Dec. 10–13, 1973, Goddard.

29. The first of two key innovations, an algorithm called *cubic convolution resampling* invented at TRW, was used for geometric corrections. The cubic convolution method efficiently transformed the distorted matrix that resulted from applying the geometric corrections to the regular matrix of pixels (picture elements) making up the image into a new undistorted matrix without losing information. The second was the development of a fast, reliable version of a special-purpose computer called a *pipeline* or *array processor* (invented in the early 1970s for the Department of Defense). This computer performed a series of operations on a series of data points at the same time by moving each one one step down the pipeline of operations each time a new one was put in the top. This was faster than a general-purpose computer, which moves each data point from memory to the buffer, performs the operations on it, and moves it back to memory, one point at a time.

30. Charles F. Mathews to John F. Clark, Robert N. Lindley, and Stanley Weiland, "NASA Data Processing Facility Augmentation," Dec. 7, 1973; "ERTS NDPF" folder, Information Processing Division, Goddard.

31. J. H. Boeckel to C. R. Laughlin, "NDPF Modifications: OSIP Task Description," Sept. 10, 1973; Goddard Space Flight Center, "NASA Data Processing Facility Augmentation: Mini Project Plan," January 1974; both in "ERTS NDPF" folder, Information Processing Division, Goddard. John F. Clark to Mr. Truszynski, "Categorization of a Minicomputer Controller for the ERTS Digital Image Pre-Processing System (ADP-592, Part of A-83 Hardware System 5103-BB)," Feb. 7, 1974; "ERS, ERTS" folder, Information Processing Division, Goddard. Interview with William C. Webb, Goddard, July 17, 1979.

Chapter 10

1. In *Science and Action* (Cambridge: Harvard Univ. Press, 1987), pp. 132–136, Bruno Latour criticizes the diffusion model of how a new technology comes into use. I use the term diffusion not in the sense of a model of passive diffusion but as a general term for the process by which promoters and customers work out the spread of a technological innovation.

2. For one approach to a model that deals with this complexity see Thomas Parke Hughes, *Networks of Power: Electrification in Western Society. 1880–1930* (Baltimore: Johns Hopkins Univ. Press, 1983).

3. Bruno Latour analyzes the importance of the process of bringing a technological innovation into use in *Science in Action*, pp. 132–144. Note that my discussions of user redefinition are not cases of passive resistance to a new idea, a view that Latour argues is too simple, or the shifting alliances around an innovation that Latour emphasizes in his discussion of the translation of an innovation into practical use. Rather, the case of Landsat showed interest groups promoting competing definitions of a satellite system even after the satellite was launched because the system could be redefined to meet various interests by reshaping the use of the data as completely as it could be redefined by hardware design.

4. M. R. Holter to L. Jaffe, "Aircraft Program Comments," no date, but routing slip dated Nov. 9, 1970; History Office, Johnson Space Center, Houston, Tex.

5. William Nordberg to multiple addressees, "Minutes of ERTS User Agency Meeting—13 March 1970 " Apr. 14, 1970; Information Processing Division files, ERTS Users Working Group, Goddard Space Flight Center, Greenbelt, Md. Goddard Space Flight Center, "Design Study Specifications for the Earth Resources Technology Satellite: *ERTS-A* and *-B*," revision B, S-107-P-3, Jan. 1970; WNRC 255-77-395, box 1. Ad Hóc Group on ERS Data Management, "Earth Resources Technology Satellite (ERTS) Data Management," Feb. 1970; WNRC 255-77-269, box 2.

6. Interviews with John DeNoyer, U.S. Geological Survey, Reston, VA, Aug. 7, 1979, and with Thomas Ragland, Goddard, Aug. 2, 1979. John M. DeNoyer to the Deputy Associate Administrator (Science), March 16, 1970; History Office, "Documentation—Earth Resources," NASA.

7. Interview with Stanley Freden, Goddard, Aug. 2, 1970. Harold H. Levy to multiple addressees, Progress Report for ERTS for the Week Ending November 2,

1969," Nov. 3, 1969; Information processing Division files, ERTS Weekly Progress Report (Levy) 12/67-10/72," Goddard.

8. Robert F. Allnut to Thomas G. Abernathy, Aug. 12, 1969; History Office, "Documentation—Earth Resources," NASA.

9. Marvin F. Matthews to distribution, "NPD 8000.1 and Dissemination of Earth Resources Photographic Data," July 27, 1972; History Office, Johnson. Christopher C. Kraft, Jr., to Charles W. Mathews, Oct. 17, 1972; History Office, Johnson.

10. Stanley C. Freden and Enrico P. Mercanti, eds., *Symposium on Significant Results Obtained from Earth Resources Technology Satellite-1, Volume 3: Discipline Summary Reports* (Greenbelt, Md.: Goddard Space Flight Center X-650-73-155, May 1973).

11. O. Glenn Smith to Chief, Earth Observations Division, "General Observations from Recent ERTS Investigation Assessment Pertinent to EREP Investigations," Jan. 15, 1974; History Office, Johnson.

12. C. E. Charlesworth to Director, Earth Observation Programs, "ERTS Follow-on Investigations," 20 Sept. 1974; History Office, Johnson. O. Glenn Smith telex to J. Boeckel and S. Freden, "Analysis of ERTS Follow-on Investigations," 8 March 1974; History Office, Johnson.

13. Charlesworth to Director, Sept. 20, 1974.

14. Stanley C. Freden to J. Morrison, "Justification for Continuing Support of a Landsat Investigator Program," April 25, 1975; History Office, Johnson.

15. Vincent V. Salomonson, "Water Resources," in Freden and Mercanti, *Symposium on Significant Results*, vol. 3.

16. C. E. Charlesworth to Mission Utilization Office, Goddard, "Scientific Monitor for the Texas ERTS-1 Follow-on Investigation," Oct. 17, 1974; History Office, Johnson.

17. C. E. Charlesworth to J. W. Jarman, March 4, 1974; History Office, Johnson.

18. John P. Erlandson to R. B. MacDonald, March 16, 1976; History Office, Johnson.

19. Interview with John W. Jarman, U.S. Army Corps of Engineers, Washington, D.C., July 31, 1979.

20. James R. Greaves, "Marine Resources and Ocean Surveys" in Freden and Mercanti, *Symposium on Significant Results*, vol. 3.

21. Interview with Polcyn. Alden P. Colvocoresses to the record, "NASA/Cousteau Ocean Bathymetry Experiment," Feb. 13, 1976; H. Mannheimer's files, "*Landsat-C* General 1971–76," NASA.

22. Alden P. Colvocoresses to the record, "Developments in Shallow-seas Mapping Application of Landsat," Nov. 4, 1976; H. Mannheimer's files, "DOI/USGS Memorandum for Record (EC-)," NASA.

23. Alden P. Colvocoresses to the record, "Plans for High-Gain Landsat Coverage of Coastlines and Shallow Seas," Sept. 1, 1978: H. Mannheimer's files, "DOI-USGS Memorandum for Record (EC-)," NASA.

24. Jarman interview.

Chapter 11

1. "Final Report: Ground Data-Handling System Study for the Earth Resources Observation Satellite: Phase 1," RCA Astro Electronics Division report #R-3408, Dec. 15, 1968.

2. William A. Radlinski, "Meeting on EROS Reception Center: October 16, 1969;" Glenn Landis's files, "Prog. 3.1," EROS Data Center.

3. David L. Stenseth, "EROS—The Local Story," *IDEL Earth Trak* 1 (June 1974): 4–5.

4. Interview with R. A. Pohl, EROS Data Center, Sioux Falls, S. Dak., May 5, 1981. Conversation with David L. Stenseth, Sioux Falls, S. Dak., May 6, 1981.

5. "News Release from the Office of U.S. Senator Karl E. Mundt," March 30, 1970: Glen Landis's files, "Prog. 3.1," EROS Data Center. Stenseth, "EROS—The Local Story."

6. Carl T. Curtis to Walter J. Hickel, June 29, 1970; Glen Landis's files, "Prog. 3.1.," EROS Data Center.

7. R. A. Pohl to William A. Fischer, Nov. 18, 1969; R. A. Pohl's files, EROS Data Center.

8. "Industrial and Development Foundation Board of Directors: Minutes," Aug. 19, 1971; R. A. Pohl's files, EROS Data Center.

9. Interview with James R. McCord, EROS Data Center, May 7, 1981.

10. John M. DeNoyer to the Associate Administrator for Space Science and Applications, "Background Material for October 16 ERSPRC Meeting," Oct. 12, 1970; History Office, "Earth Resources Documentation," NASA.

11. Interview with Fredericka A. Simon, EROS Data Center, May 6, 1981.

12. Interview with Allen H. Watkins, EROS Data Center, May 8, 1981.

13. Watkins interview.

14. Robert L. Bell to Robert O. Piland, Oct. 22, 1970; History Office, "*Landsat 1* Documentation," NASA.

15. Homer E. Newell to Donald B. Rice, July 22, 1971; David Williamson's files, "ERSPRC—July 8, 1971—Correspondence," NASA.

16. Interview with Donald F. Carney, EROS Data Center, May 6, 1981.

17. Thomas M. Ragland to EROS Program Manager, "EDC Color Composite and Other Photographic Consideration," Aug. 26, 1976; Office of the Chief, "GSFC/EDC Interface," EROS Data Center.

18. Winston Sibert to Associate Director, "The Role of the EROS Data Center in Application of Remote Sensing Data to Benefit Energy and Mineral Resources Activities," May 1, 1974; Office of the Chief, "Reston Interface," EROS Data Center.

19. Allen H. Watkins to Frederick J. Doyle, Feb. 21, 1978; Office of the Chief, "Reston," EROS Data Center.

20. In September 1976 the EROS Data Center staff consisted of 50 civil servants and 275 contractor personnel.

21. William B. Gevarter, "Minutes of the ERS Data Management. Working Group (DMWG) Meeting at GSFC, July 23, 1970;" WNRC 255-81-416, box 9, #004696.

22. William R. Hemphill to the Record, "Trip Report— Collaborative EROS Program Activities with Bureaus of Reclamation and Land Management/Denver," July 7, 1976; Office of the Chief, "Reston," EROS Data Center. The concept of takeoff came from a theory of technology transfer and economic development that maintained that beyond a certain point of development, economic growth would become self-sustaining (W. W. Rostow, *The Stages of Economic Growth* (Cambridge: Cambridge Univ. Press, 1960)). By analogy the use of a new tool such as Landsat data had to be heavily promoted to each user until a certain momentum was achieved in an organization, and then adoption would continue to occur by itself.

23. Allen H. Watkins to Allan Marmelstein, 20 Aug. 1976; Office of the Chief, "Reston," EROS Data Center.

24. Gary G. Metz to William R. Hemphill, Feb. 17, 1978; Office of the Chief, "Reston," EROS Data Center.

25. "Statement of Cecil D. Andrus, Secretary of the Interior, for the Subcommittee on Science and Space, Committee on Commerce, Science, and Transportation, U.S. Senate," March 15, 1977; EROS Program Office, Significant Documents file.

26. "The International Training Program of the EROS Data Center," March 1981; William C. Draeger's files, EROS Data Center.

27. "User Assistance and Training," Oct. 7, 1978: Office of the Chief, "Reston," EROS Data Center.

28. Gary G. Metz to William R. Hemphill, Feb. 17, 1978; Office of the Chief, "Reston," EROS Data Center.

Chapter 12

1. Homer E. Newell to Stanley M. Greenfield, June 30, 1971; History Office, Johnson Space Center, Houston, Tex.

2. "246 Counties Will Use Aerial Photos to Spot Corn Blight," *The Sioux City Journal Farm Weekly*, Apr. 19, 1971; Scrapbook, EROS Data Center.

3. Robert N. Colwell, *Monitoring Earth Resources from Aircraft and Spacecraft* (Washington, D.C.: NASA SP-275, 1971).

4. "Report Given on SRS Remote Sensing Research for Top USDA and NASA Management," attachment to Donald H. Von Steen to A. W. Fihelly, Oct. 15, 1973; "Investigator #1013" folder, box 1, accession number 79-283, record group 255, Washington National Records Center.

5. A. E. Potter to Arch P. Park, facsimile transmission, "1972 Input to COSPAR Report," Dec. 21, 1972; History Office, Johnson.

6. Interview with Robert B. MacDonald, Johnson, Sept. 10, 1979.

7. John M. DeNoyer to the Deputy Administrator and the Associate Administrator, "OST Spring Preview on the Earth Resources Program," June 24, 1971; History Office, "Earth Resources Documentation," NASA.

8. "Possible ERS Questions and Answers," July 11, 1973, attachment to David Williamson, Jr., to Willis Shapley, July 18, 1973; History Office, Johnson. Willis Shapley to distribution, "NASA Posture on Crop Forecasts," July 10, 1973; History Office, Johnson.

9. MacDonald interview.

10. Charles W. Mathews to the record, "World-Wide Wheat Survey Demonstration Program," Aug. 30, 1973; Bryan Erb's files, "Management and Interagency Reviews," Johnson.

11. John D. Overton to distribution, "Project Management Planning Activity," Oct. 19, 1973; Bryan Erb's files, "LACIP Early Documentation," Johnson. James C. Fletcher to John C. Sawhill, Oct. 19, 1973; H. Mannheimer's files, "OMB Budget Justification," NASA.

12. C. E. Charlesworth to NASA Hq: Director, Earth Observations Programs, "Large Area Crop Inventory Project (LACIP) Key Issues " [Feb. 21, 1974]; Bryan Erb's files, "LACIE Historical, Johnson.

13. Charles W. Mathews to the Administrator, "Large Area Crop Inventory Experiment (LACIE)," Sept. 26, 1974; History Office, Johnson. R. B. MacDonald to Distribution, "LACIE Interagency Project Review at JSC, October 7–8, 1974," Nov. 7, 1974; Bryan Erb's files, "1.1 ERPO," Johnson.

14. Interview with Bobby Spiers, Foreign Agricultural Service, Houston, Tex., March 19, 1981.

15. James E. Powers to distribution, "LACIE Agency Project Managers Meeting, August 29, 1975," Oct. 17, 1975; History Office, Johnson.

16. C. E. Charlesworth to Distribution, "Minutes of the LACIP Review on August 14, 1974, for the Associate Administrator of the Office of Applications and Staff, Initiated and Presented by the Earth Resources Program Manager LACIP Project Manager, and their Staffs, Respectively," Aug. 20, 1974; Bryan Erb's files, "1.1 ERPO," Johnson.

17. R. B. MacDonald to Manager, Earth Resources Program, "Minutes of LACIP Review on August 14, 1974, for Associate Administrator, Office of Applications and Manager, Earth Resources Program," Sept. 13, 1974; Bryan Erb's files, "1.1 ERPO," Johnson.

18. Interview with Jim Hickman and Andy Aaronson, Foreign Agricultural Service, March 18, 1981. Interview with Del Conte, Agristars Program, Houston, Tex., March 19, 1981.

19. Howard W. Tindall, Jr., to Director of Science and Applications, Johnson, "Large Area Crop Inventory Project (LACIP) Support," May 2, 1974; History Office, Johnson.

20. "Large Area Crop Inventory Experiment Management Report #7 for the Period August 1975," approved by R. B. MacDonald, Sept. 19, 1975; Bryan Erb's files, "Bimonthly Management Reports," Johnson.

21. R. B. MacDonald to Director, Earth Observations Programs, "Response to Fletcher Questions," Aug. 19, 1975; History Office, Johnson.

22. "Large Area Crop Inventory Experiment Management Report #9 for the Period November 1975," approved by R. B. MacDonald, Jan. 13, 1976; Bryan Erb's files, "Bimonthly Management Reports," Johnson.

23. Charles W. Mathews to the Administrator, "Technical Review of LACIE Operating Procedure," Aug. 27, 1975; Bryan Erb's files, "1.11 OA Correspondence," Johnson.

24. T. L. Branett and W. J. Crea to the record, "Report on Travel to Kansas State University, June 11–12, 1975," June 18, 1975; History Office, Johnson.

25. Robert MacDonald to Director, Earth Observation Programs, "Status of Remote Sensing Crop Identification Accuracies in Relation to LACIE Performance Requirements," [no date]; Bryan Erb's files, "1.2a NASA Hq. 1974," Johnson.

26. The sources of this increased emphasis on cost-benefit analysis were complex. Chapter 7 gives background, and a number of studies of the space shuttle have touched on the broader issue. See Alex Roland, "The Shuttle: Triumph or Turkey?" *Discover* 6 (Nov. 1985): 29–49, and John M. Logsdon, "The Space Shuttle Program: A Policy Failure?" *Science* 232 (May 30, 1986): 1099-1105.

27. Michael A. Calabrese to distribution, "Minutes of the *Landsat-D* Mission Planning Meeting at GSFC on September 4, 1975," Sept. 9, 1975; History Office, Johnson. Craig L. Wiegand to William Nordberg and Leonard Jaffe, "*Landsat-D* Launch Uncertainty," Sept. 10, 1975; Bryan Erb's files, "1.3 USDA," Johnson.

28. William H. Anderson to the record, "Trip Report Concerning LACIE Project Phase 11 Interim Review, January 26–28, 1977 at NASA JSC," Feb. 25, 1977: Office of the Chief, "Applications Branch," EROS Data Center.

29. R. B. MacDonald to distribution, "Key Project Issues," March 22, 1976; Bryan Erb's files, "1.2 NASA Hq 1976," Johnson.

30. Eliot Cutler to Howard Hjort, May 19, 1977; Bryan Erb's files, "1.13 OMB," Johnson.

31. R. B. MacDonald to Distribution, "Review of the LACIE Phase III Results—A Response to the SRS Evaluation of LACIE," Oct. 21, 1977; Bryan Erb's files, "1.2 NASA Hq 1977," Johnson.

32. J. Hill to Manager, LACIE Project Office, "Phase III Activities of the LACIE Crop Conditions Assessment Team," Dec. 23, 1977; Bryan Erb's files, "1.4 NOAA," Johnson.

33. "Comptroller General's Report to the Congress. Crop Forecasting by Satellite: Progress and Problems. Digest," attachment to Walter C. Shupe to distribution, "GAO Report to Congress 'Crop Forecasting by Satellite: Progress and Problems,' B-183184, April 7, 1978," Apr. 21, 1978; History Office, Johnson.

34. R. B. MacDonald to Distribution, "LACIE Review of the Acceptability of LACIE Methodology," May 21, 1976: Bryan Erb's files, "1.2 NASA Hq 1976," Johnson.

35. Interview with Barry H. Moore, Agristars, March 20, 1981.

36. W. H. Anderson to the Record, "LACIE Review," July 10, 1978; Office of the Chief, "Applications Branch," EROS Data Center.

37. James L. Mitchell to Richard E. Bell and Don Paarlberg, "Large Area Crop Inventory Experiment," Oct. 8, 1976; Bryan Erb's files, "1.13 OMB," Johnson. Cost-benefit calculations were tricky because the effects of improved crop forecasting accrued to farmers and grain dealers through complicated processes in the commodity market.

38. John Logsdon, "The Crop Assessment Program of the Foreign Agricultural Service: A Case Study of Applied Space Technology," unpublished paper.

39. Interview with Bryan Erb, Johnson, Sept. 13, 1979.

40. Hickman and Aaronson interview.

41. Ibid.

42. Logsdon, "The Crop Assessment Program."

43. Conte interview.

Chapter 13

1. David W. Jamison, "Earth Resources Observation Satellite Image Utility at State Level," April 1970, USGS Grant #14-08-0001-G-21; on file in the EROS Data Center Library, Sioux Falls, S. Dak.

2. George M. Low to NASA Center Directors, "Earth Resources Survey Program," Nov. 19, 1970; R. Weinstein's files, NASA.

3. Sigurd A. Sjoberg to Harry H. Gorman, April 8, 1972; History Office, Johnson Space Center, Houston, Tex.

4. Paul R. Kramer to Christopher Kraft, Sept. 18, 1972; History Office, Johnson.

5. Alden P. Colvocoresses to Leonard Jaffee, "Initial Space Flight Objective of Earth Resources Survey Program." Feb. 28, 1967; History Office, "Documentation—Earth Resources," NASA.

6. James C. Fletcher to Frank E. Moss, Feb. 20, 1973; History Office, "Documentation—Earth Resources," NASA.

7. Clinton P. Anderson to James C. Fletcher, Oct. 14, 1972; History Office, "Fletcher Correspondence," NASA.

8. Christopher Kraft to Robert L. Krieger, Sept. 13, 1973; History Office, Johnson.

9. Samuel I. Doctors, *The NASA Technology Transfer Program: An Evaluation of the Dissemination System* (New York: Praeger Publishers, 1971), pp. 6, 53, 159.

10. Granville W. Hough, *Technology Diffusion: Federal Programs and Procedures* (Mt. Airy, Md.: Lombard Books, 1975), pp. 190–191.

11. Ibid., 225.

12. James C. Fletcher to John C. Sawhill, Oct. 19, 1973; H. Mannheimer's files, "OMB Budget Justification," NASA.

13. Charles W. Mathews to distribution, "'Applications Advocacy Plan' Task Team Membership," April 16, 1974; History Office, Johnson.

14. Ibid.

15. A. T. Christenson to Members, Applications Advocacy Plan Task Team, "Task Team Activity Plan," June 10, 1974; History Office, Johnson.

16. Jim Richards to User Communication Task Team Members, "Preparation for States Visits," June 20, 1974: History Office, Johnson.

17. Herman Mark to Charles W. Mathews and Leonard Jaffe, "Interim Report on Status of Great Lakes ASVT—Project ICEWARN," Apr. 24, 1975; History Office, Johnson.

18. Ibid.

19. Ibid.

20. "Space Applications Transfer Centers," [April 7, 1976]; R. Weinstein's files, NASA.

21. Interview with Robert Piland, Johnson, Sept. 12, 1979.

22. "Johnson Space Center Activities at the Mississippi Test Facility," attachment to Christopher C. Kraft, Jr., to George M. Low, April 30, 1974; History Office, Johnson.

23. Dan Richard, "Meeting on Training and Assistance for Landsat Data Users, Goddard Space Flight Center, November 22, 1976,"; Office of the Chief, "NASA Hq," EROS Data Center.

24. Everett M. Rogers, *Diffusion of Innovations*, 1st ed. (New York: The Free Press of Glencoe, 1962). Everett M. Rogers and F. Floyd Shoemaker, *Communication of Innovations: A Cross-Cultural Approach*, 2d ed. (New York: The Free Press, 1971).

25. A. C. Robinson and J. A. Madigan, "Task Final Report on Options for Organization and Operation of Space Applications Transfer Centers," June 14, 1976 (NASA Contract NASw-2800, task 15); R. Weinstein's files, NASA, p. 1.

26. Ibid., 34.

27. Ibid.

28. Christopher C. Kraft, Jr., to Bradford Johnston, Aug. 30, 1976; R. Weinstein's files, NASA.

29. Murray Felsher to Pitt [Thome], Sept. 13, 1976; R. Weinstein's files, NASA.

30. Lee R. Scherer to Associate Administrator for Applications, "Proposed New Guidelines for Regional Applications Transfer Funding," [Sept. 1976]; R. Weinstein's files, NASA.

31. Bradford Johnston to Donald P. Hearth, Jan. 12, 1977; R. Weinstein's files, NASA.

32. "Regional Applications Program," attachment to A. Donald Goedeke to the Associate Administrator, "Restructuring of the Office of Applications' Regional Applications Program," [Mar. 2, 1977]; R. Weinstein's files, NASA.

33. Ben D. Padrick to Don A. Goedeke, draft, Feb. 8, 1977; R. Weinstein's files, NASA.

34. Murray Felsher to Deputy Director, User Affairs, "Response to Mathias's Question," Jan. 26, 1977; R. Weinstein's files, NASA. A. Donald Goedeke to the Deputy Administrator, "Wallops Flight Center Regional Applications Status Update," April 4, 1977; R. Weinstein's files, NASA.

35. Natural Resource and Environment Task Force of the Intergovernmental Science, Engineering and Technology Panel, "State and Local Government Perspectives on a Landsat Information System," Office of Science and Technology Policy, June 1978, p. 25.

36. Ibid.

37. Ibid., XII.

38. Ibid., E-8.

39. Ibid., E-16.

40. J. A. Vitale to the Associate Administrator, "University of Nebraska—OUA Space Applications Program," [Sept. 17, 1976]; History Office, Earth Resources Documentation, NASA.

Chapter 14

1. Martin W. Molloy to Chief, Earth Resources Survey Program and Program Scientists, ERTS and Skylab EREP, "ERTS Results Relating to Petroleum Exploration," Oct. 4, 1973; History Office, Johnson Space Center, Houston, Tex.

2. Christopher C. Kraft, Jr., to James C. Fletcher, Dec. 13, 1973; History Office, Johnson.

3. Christopher C. Kraft, Jr., to Mr. John Butler, Jr., Nov. 3, 1973; History Office, Johnson.

4. Ibid. and Kraft to Fletcher, Dec. 13, 1973.

5. Milford R. Lee to Anthony Calio, 7 Nov. 1973; History Office, Johnson.

6. F. C. Canny to Larry Rowan, "Review of 'The Thing,'" Nov. 5, 1973; History Office, Johnson.

7. Frederick B. Henderson, talk to a meeting sponsored by the Geosat Committee, Jan. 27, 1981, Department of the Interior, Washington, D.C.

8. Martin W. Molloy to Gene A. Thorley, "Discussion of Petroleum Exploration Research," Dec. 20, 1973; History Office, Johnson.

9. Martin W. Molloy to the record, "Meeting Report— Requirements and Benefits Subcommittee ICCERSP—March 14, 1973," May 4, 1973; History Office, Johnson.

10. John M. DeNoyer to T. A. George, March 25, 1974; Office of the Chief, "Statistics," EROS Data Center, Sioux Falls, S. Dak.

11. G. M. Low, note to Dr. Newell, Jan. 14, 1974: History Office, "Fletcher Correspondence," NASA.

12. Martin W. Molloy to Director, Earth Observations Programs and Manager, Earth Resources Survey Programs, "Potential ASVT with USGS," Aug. 13, 1975; History Office, Johnson.

13. D. O. Nelson to Marty Molloy and Ambionics, Inc., Aug. 7, 1975; History Office, Johnson.

14. D. O. Nelson to Marty Molloy, Aug. 8, 1975; History Office, Johnson.

15. Untitled sheet with the control number IC 12-153 and a note from LEO to A. W[atkins], Dec. 22, 1975; Office of the Chief, "Statistics," EROS Data Center.

16. Dr. Leroy Scharon, Director of Exploration, National Lead Industries, quoted in EROS Program, "Implementation of an Operational Earth Resources Satellite System," Feb. 1977; Significant Documents file, EROS Program Office, Reston, Va.

17. Interview with J. Robert Porter, Jr., Earthsat Corp., Bethesda, Md., July 27, 1978.

18. "Functional Requirements of Landsat Follow-On (non-DOI)," June 20, 1976: Office of the Chief, "ERTS/Landsat," EROS Data Center.

19. Applications Branch, EROS Data Center, "Sources for Image Processing and Interpretation Support," May 1978; Office of the Chief, "Applications Branch," EROS Data Center.

20. "Remote Sensing at Environmental Research Institute of Michigan," pamphlet [1973].

21. Interview with Fabian Polcyn, Environmental Research Institute of Michigan, Ann Arbor, Mich., May 11, 1981.

242   Notes to pages 177–184

22. Applications Branch, EROS Data Center, "Sources for Image Processing and Interpretation Support," July 1, 1977; Office of the Chief, "Reston," EROS Data Center.

23. Robert Buchanan, "Looking Down is Up," *Washington Star*, March 4, 1973. "Statement of J. Robert Porter, Jr., President, Earth Satellite Corporation, Before the House Subcommittee on Space Science and Applications: June 23, 1977."

24. Porter statement, June 23, 1977.

25. Robin I. Welch to William Nelson, Dec. 28, 1973; History Office, Johnson.

26. Porter statement, June 23, 1977.

27. Ibid.

28. In its own defense the EROS Data Center pointed to some industry statements that supported the role of the center. For example, the director of the EROS data center reported that Gene Zaitzeff of the Bendix Corporation said that "as an industrial processor of CCT [computer compatible tape] data their company has benefited directly from the training given at EDC through the increased awareness of the customer community regarding the capabilities of this technique." Allen H. Watkins to William A. Fischer, Feb. 26, 1976; Office of the Chief, "Reston," EROS Data Center. See also Robert K. Vincent to William Fischer, Jan. 28, 1976; Office of the Chief, Reston," EROS Data Center.

29. Christopher C. Kraft, Jr., to Harry H. Gorman, May 1, 1974; History Office, Johnson.

30. William A. Fischer to Robert K. Vincent, March 9, 1976; Office of the Chief, "Reston," EROS Data Center.

31. Arnold W. Frutkin to Dr. Newell, "Proprietary ERS Data," May 8, 1970, with attachment: Leonard Jaffe for the Record, "Visit by General Electric Company Representatives March 25, 1970, Re: Commercial Exploitation of Earth Resources Survey Satellites," 30 March 1970; WNRC: 255-77-0677, box 33, "Related Sciences 3, 1970, ER/ERS/EROS/ERTS—Remote Sensing."

32. "A Discussion of the Possible Role of Industry in the Landsat System," attachment to Leonard Jaffe to John DeNoyer, Sept. 10, 1976; EROS Program dead files, "NASA—General (1973- 76)."

33. Unclassified form of Presidential Directive PD-37, attachment to John M. DeNoyer to multiple addressees, Significant Documents Related to Remote Sensing From Space," July 3, 1978; EROS Program Office, "Significant Documents."

Chapter 15

1. For an excellent introduction to the history of space treaties, see Harold M. White, Jr., "International System and Space Law: An Introduction," in T. Stephen Cheston, Charles M. Chafer, and Sallie Birket Chafer, *Social Sciences and Space Exploration: New Directions for University Instruction* (Washington D.C.: NASA EP-192, 1984).

2. Willis B. Foster to Assistant Administrator for International Affairs, "Remote Sensing Project Test Sites outside of the United States," Apr. 22, 1965; "SM Reading File 1965" folder, box 52, accession number 74-663, record group 255, Washington National Records Center.

3. The same argument was made for a wide range of international projects established by a wide range of organizations. For example, the Purdue University School of Agriculture had an Agency for International Development project in the same area of Brazil that could provide support for remote sensing work. For both cases see Willis B. Foster to the Assistant Administrator, "Foreign Test Sites," June 25, 1965; "SM Reading File 1965," 255-74-663 (52), WNRC.

4. Another practical advantage was the possibility that cooperative research on locating minerals might be more easily conducted with other countries than with private industry (see chapter 14). On the other hand, the Department of the Interior had used international cooperation as a tool in interagency competition for influence over Landsat from the beginning (see chapter 5).

5. R. R. Gilruth to Leonard Jaffe "Earth Resources International Participation Program," Sept. 3, 1968; History Office, Johnson Space Center, Houston, Tex.

6. Christopher C. Kraft, Jr., to Charles W. Mathews, July 27, 1972; History Office, Johnson.

7. For an excellent discussion of these issues in more detail than is possible here see *Remote Sensing and the Private Sector: Issues for Discussion—A Technical Memorandum*, (Washington D.C.: U.S. Congress, Office of Technology Assessment, OTA-TM-ISC-20, Mar. 1984).

8. For a good discussion of the development of open skies policy for reconnaissance satellites see Walter McDougall, . . . *the Heavens and the Earth: A Political History of the Space Age* (New York: Basic Books, 1985).

9. Edward Ezell and Linda Ezell, *The Partnership: A History of the Apollo-Soyuz Test Project* (Washington, D.C.: NASA SP-4209, 1978), p. 27.

10. A. W. Frutkin to Dr. Fletcher, "International Evolution of ERS Systems," Jan. 5, 1972; History Office, "Fletcher Correspondence," NASA.

11. Ibid.

12. Ibid.

13. Ibid.

14. A 1978 statement suggested a limiting resolution of fifty yards. "USSR Wants 50-yd Resolution Limit on Civil Space Photographs," *Defense/Space Daily* (Feb. 28, 1978): 310.

15. "Policy Relating to ERTS and EREP Experiments," Nov. 23, 1971, attachment to A. W. Frutkin to Dr. Fletcher, Jan. 5, 1972; History Office, "Fletcher Correspondence," NASA.

16. For a summary of such proposals see John H. McElroy, Jennifer Clapp, and Joan C. Hock, "Earth Observations: Technology, Economics, and International Cooperation," a background document for the National Academy of Engineering and Resources for the Future Symposium, "Explorations in Space Policy: Emerging Economic and Technical Issues," Washington, D.C., June 24–25, 1986, particularly p. I-23.

17. B. Cheng, "Legal Implications of Remote Sensing from Space," *Proceedings of an International Conference on Earth Observation from Space and Management of Planetary Resources*, held at Toulouse, 6–11 March 1978 (ESA SP-134).

18. Quoted in Remote Sensing and the Private Sector, p. 41.

19. Ibid., 103.

20. "Latin America Alters View on Landsat," *Aviation Week and Space Technology* (July 17, 1978): 16–17.

21. Arnold W. Frutkin to Dr. Fletcher and others, "Some Recent International Reactions to ERTS-1," Dec. 22, 1972; History Office, "Fletcher Correspondence," NASA.

22. U.S. Embassy, Bamako, Mali, State Department Telegram, Nov. 17, 1972, quoted in Frutkin to Fletcher, Dec. 22, 1972.

23. See *Remote Sensing and the Private Sector*, Appendix A, for examples.

24. Interview with Murray Strome of the Canadian Centre for Remote Sensing at the International Symposium on Remote Sensing of Environment, Ann Arbor, Mich., May 12, 1981.

25. Thomas M. Ragland to Distribution, "Cooperation With Canadians," Nov. 4, 1970; #005440, 255-81-416 (9), WNRC.

26. Strome interview.

27. Martin W. Molloy to the Record, "USDI/EROS Staff Meeting—December 7, 1972," Jan. 26, 1973; History Office, Johnson.

28. Strome interview.

29. This figure was increased to $600,000 per year in 1982.

30. EROS Data Center, "A Summary of Worldwide Landsat Data Demand for Calendar Years 1979–80," prepared for the International User Services/Data Management Workshop in Canberra, Australia, May 18–19, 1981, revised June 1, 1981; R. A. Pohl's files, EROS Data Center. By 1985, additional stations in Brazil and Italy were operating, and others in China, Indonesia, Pakistan, Saudia Arabia, and Thailand were in various stages of development. *International Cooperation and Competition in Civilian Space Activities* (Washington D.C.: U.S. Congress, Office of Technology Assessment, OTA-ISC-239, July 1985), p. 284.

31. Adigun Ade Abiodun, "The Economic Implications of Remote Sensing from Space for the Developing Countries," *Proceedings of an International Conference on Earth Observation From Space and Management of Planetary Resources held at Toulouse, 6–11 March 1978* (ESA SP-134).

32. "A Summary of Worldwide Landsat Data Demand for Calendar Years 1979–80."

33. Leonard Jaffe to the record, "Meeting with USSR Members of the US/USSR Joint Working Group on the Natural Environment," Oct. 25, 1973; History Office, Johnson.

34. "Scientist Writes on Benefits of In-Orbit Photography," *Pravda* Aug. 14, 1979, translated in "21 Aug. 1979, USSR National Affairs: Scientific Affairs;" History Office, "Fletcher Correspondence," NASA. "Soviets Launch Landsat, NASA Daily Activities Report, June 20, 1980.

35. Pierre Langereux, "France Puts Landsat on the Spot," *Aerospace America* 24 (May 1986): 8–9.

36. "SPOT: Satellite-Based Remote Sensing System," pamphlet published by the Centre National d'Etudes Spatiales and distributed at the Fifteenth International Symposium on Remote Sensing of Environment, May 10–15, 1981, Ann Arbor, Mich.

37. "Japan Plans Resource Observation Satellite System," *Aerospace Daily* (11 July 1979): 55.

38. "NOAA Task Force Looks to Middle Ground in Remote Sensing Assessment," *Aerospace Daily* (Aug. 11, 1980): 229.

Chapter 16

1. This was not an easy matter in the tight-budget times of the late 1960s and 1970s. For a helpful theoretical discussion of agency strategies see W. Henry Lambright, *Presidential Management of Science and Technology: The Johnson Presidency* (Austin: Univ. of Texas Press, 1985).

2. "Discussion Paper," Apr. 30, 1970, attachment to Dudley G. McConnell to the Record, "Earth Resources Program: NASA-BoB Meeting on Proposed BoB Organization and Management Study," May 5, 1970; "Earth Resources Survey Program Review Committee, (ERSPRC)," box 1, accession number 75-704, record group 255, Washington National Records Center.

3. For specific examples of this role see chapters 6 and 8.

4. D. Williamson, "Policy Paper," draft May 11, 1971, attachment to Homer E. Newell to the Administrator, "Statements of Policy with Respect to NASA's Role in Space Applications," May 18, 1971; Leonard Jaffe's chronological file, NASA.

5. Charles W. Mathews to John F. Clark, "Guidelines for Earth Observatory Satellite (EOS) Program," June 23, 1972; History Office, Johnson Space Center, Houston Tex.

6. "NASA Position Paper on the Earth Resources Survey Program," Nov. 1970, p. 17; "Related Sciences 3, ER, ERS, ERTS 1971–1972," 255-77-0677 (33), WNRC. In fact, arguments over data specifications (see chapter 6) had already proved that user needs affected the space portion of the system—the users were unlikely to be satisfied by a generic data collection system whose design was decided by the space agency.

7. "Summary of Economic Benefit Studies," Appendix A of a draft white paper, "Earth Resources Technology Satellite Program," Nov. 17, 1968; D. Williamson's files, "ERSPRC, 1968, Dr. Naugle," NASA.

8. Advanced Programs Office, Science and Applications Directorate, "Statement of Work: Phase A Definition of an Operational Earth Resources Survey System," July 21, 1972; History Office, Johnson.

9. Homer E. Newell to the record, "Interagency Coordination Committee for the Earth Resources Survey Program Federal Report and Plan," Sept. 25, 1972; D. Williamson's files, "Minutes of the ICCERSP," vol. 1, NASA.

10. C. E. Charlesworth to Acting Director, Earth Observations Program, NASA Headquarters, "Earth Resources Survey Operational System," Nov. 7, 1972; History Office, Johnson.

11. The improvements in data processing corrected inadequacies in the main data distribution system (see chapter 9), while the new infrared band was selected to be particularly useful for agricultural applications.

12. W. Nordberg to C. W. Mathews and W. E. Stoney, "'5th Band MSS'—EOS Mission," Feb. 13, 1974; History Office, Johnson.

13. Christopher C. Joyner and Douglas R. Miller, "Selling Satellites: The Commercialization of LANDSAT," *Harvard International Law Journal* 26 (Winter 1985): 67.

14. Thomas M. Ragland to EROS Program Manager, "Information Requirements for an Operational Earth Resources System," Nov. 12, 1973; Office of the Chief, "GSFC/EDC Interface," EROS Data Center, Sioux Falls, S. Dak.

15. "Department of Agriculture Response to Question Asked by Senate Committee on Aeronautics and Space Science, February 5, 1976," attachment to Don Paarlberg to Frank E. Moss, March 2, 1976; History Office, "Fletcher Correspondence," NASA. See chapter 12 to put this in context.

16. See chapter 13 for a discussion of technology transfer for Landsat.

17. Arthur G. Anderson to James C. Fletcher, May 27, 1976; History Office, "Fletcher Correspondence," NASA.

18. William E. Stoney to Dr. W. Nordberg, "Review Meeting of Major Planned Earth Observations Missions," Apr. 18, 1974; History Office, Johnson.

19. C. E. Charlesworth to Director, Earth Observations Program, "Response to DOI Letter on Landsat Follow-on," Apr. 2, 1976; History Office, Johnson.

20. H. S. Moore to the record, "IDT Working Group Meeting 3.24.77," March 29, 1977; H. Mannheimer's files, "(IDT) Interagency Decision Team Working Group Meeting," NASA.

21. Donald P. Rogers, "Minutes: IDT Working Group Meeting NASA Headquarters, May 31, 1977;" H. Mannheimer's files, "(IDT) Interagency Decision Team Working Group Meeting," NASA.

22. M. Rupert Cutler, USDA, to Bert Lance, OMB, "Landsat Follow-on Program (NASA)," July 28, 1977; B. Erb's files, "1.13 OMB," Johnson.

23. Michael A. Calabrese to distribution, "Minutes of the Interagency Decision Team Working Group (IDTWG) Meeting of August 11, 1977," Aug. 29, 1977; H. Mannheimer's files, "(IDT) Interagency Decision Team Working Group Meeting," NASA.

24. Ibid.

25. W. Radlinski to Secretary of the Interior, "Implementation of an Operational Earth Resources Satellite System," Mar. 22, 1977; EROS Program Office, "Significant Documents."

26. Bruno Augenstein, Willis H. Shapley, and Eugene B. Skolnikoff, "Earth Information from Space by Remote Sensing," report prepared for Dr. Frank Press, Director, Office of Science and Technology Policy, June 2, 1978.

27. "Recommendations of the National Governor's Association, National Conference of State Legislatures, Intergovernmental Science, Engineering & Technology Advisory Panel, Natural Resources & Environment Task Force for the Final Transition Plan for the National Civil Operating Remote Sensing Program," Apr. 30, 1980, attachment to Richard D. Lamm to George S. Benton, Apr. 30, 1980; R. Weinstein's files, NASA.

28. "United States Space Policy: Announcement of the President's Decisions Concerning Land Remote-Sensing Activities, November 20, 1979," *Weekly Compilation of Presidential Documents* Monday, Nov. 26, 1979, vol. 15, number 47, pp. 2149–2150. Most of the discussion of earth resources policy in PD-54 grew out of a call for action in Presidential Directive 42, issued Oct. 10, 1978. For one point of

view on the evolution of PD-42 and the unclassified text see Hans Mark, *The Space Station: A Personal Journey* (Durham, N.C.: Duke Univ. Press, 1987), pp. 77–80 and 230–234.

29. M. Mitchell Waldrop, "Imaging the Earth (II): the Politics of Landsat," *Science* 216 (Apr. 2, 1982): 40–41.

30. "Recommendations of the National Governor's Association," April 30, 1980.

31. For an analysis of the proper government role in earth resources satellites on the basis of a division between public goods and private goods see John H. McElroy, Jennifer Clapp, and Hoan C. Hock, "Earth Observations: Technology, Economics, and International Cooperation," a background document for the National Academy of Engineering and Resources for the Future Symposium "Explorations in Space Policy: Emerging Economic and Technical Issues," Washington, D.C., Jan. 24–25, 1986.

32. *Effects of Users of Commercializing Landsat and Weather Satellites*, a Report by the Comptroller General of the United States (Washington, D.C.: General Accounting Office, GAO/RCED-84- 93, Feb. 24, 1984). *Remote Sensing and the Private Sector: Issues for Discussion—A Technical Memorandum* (Washington, D.C.: U.S. Congress, Office of Technology Assessment, OTA-TM-ISC-20, March 1984).

33. For a critical discussion of the policy motivations see Joyner and Miller, "Selling Satellites."

34. The story can be traced in news columns and occasional articles in *Aviation Week and Space Technology:* "Washington Roundup," (Sept. 24, 1984): 13; "Washington Roundup," (Oct. 1, 1984): 17; "News Digest," (Oct. 15, 1984): 29; "Washington Roundup," (March 25, 1985): 17; "OMB Approves Funds to Shift Landsat to Private Sector," (May 27, 1985): 19; "News Digest," (June 24, 1985): 29; "Industry Observer," (Sept. 9, 1985): 15; "Commerce Dept. Transfers Landsat Operations to Private Venture," (Oct. 14, 1985): 95–103. For an example of doubts about commercialization see Frederick B. Henderson II, "Not Defaulting on Land Remote Sensing," Aerospace America (June 1985): 28, 92.

Epilogue

1. This may be particularly significant for government-sponsored technology because the goals are more open to negotiation than in most commercial technology. Donald MacKenzie's study of inertial guidance systems for missiles seems to show the same basic factors, but the presence of a strong leader and the context of an independent laboratory result in different patterns. See Donald MacKenzie, "Missile Accuracy: A Case Study in the Social Processes of Technological Change," in Wiebe E. Bijker, Thomas P. Hughes, and Trevor Pinch, *The Social Construction of Technological Systems: New Directions in the Sociology and History of Technology* (Cambridge: MIT Press, 1987).

2. Pinch and Bijker's example of the bicycle seems to be a case where at least in the short run different definitions of a technology resulted in a split into different technologies: most notably, the high-wheeled Ordinary or Penny-farthing and the safety bicycle. Trevor J. Pinch and Wiebe E. Bijker, "The Social Construction of Facts and Artifacts: Or How the Sociology of Science and the Sociology of Technology Might Benefit Each Other," in Bijker, Hughes, and Pinch, *The Social Construction of Technological Systems.*

Notes on Sources

Archival Sources

The history of NASA is generally well documented by files kept by each office and retired when no longer current to storage at the Federal Records Centers. Landsat proved to be an exception to that pattern; I found only scattered and disorganized NASA files on Landsat. To give a sense of the strengths and weaknesses of the documentation on which this study has been based, it is necessary to describe a veritable scavenger hunt for documents.

The best place to start in almost any NASA history project is with the research collection at the NASA Headquarters History Office. Although this collection is not an official record, it represents many years of wise gleaning of magazines, internal publications, and correspondence files slated for disposal. The History Office collection contained two file drawers of material relating to Landsat and earth resources, including photographs, reports, press kits, clippings, and an interesting selection of letters and memos. Other subject files in the History Office contained material useful for project histories, particularly biography files and files of NASA Administrators.

More systematic records proved to be more difficult to find. Unlike many NASA projects, the Landsat project did not preserve a single comprehensive set of files. Some of the problems involved were suggested in a letter written by the manager of earth resources programs, T. A. George, to his superior on September 9, 1970.

As you are aware, the ERTS [Landsat] Program Office (Code SRB) has been virtually without any effective secretarial support since September 1969. The consequences of this situation are that:
(1) Other than a few key documents such as PADs, RFPs, and POPs, no useful records or files pertaining to the ERTS Program are in existence at the Headquarters level.
(2) Communication with NASA centers, contractors, and other parties interested in the ERTS Program has been largely by telephone and no record of these maintained.
(3) Incoming correspondence, action papers, and so forth have been lost.
(4) Meetings have been missed by members of the professional staff.
One of the results of this sad situation is that as of present, Headquarters would be unable to accurately reconstruct past actions and decisions. Any attempt to adequately document these events would be almost impossible.

According to NASA records the Earth Observations Program retired only fourteen boxes of records (all the boxes I looked at were 2.2 cubic foot records boxes) dealing with earth resources to the Washington National Records Center. Of those, nine contained only material on earth resources experiments on Skylab. The others contained only technical reports, except for one set of files on long-range planning.

Since most of the records of the earth resources program office have not been retired, it is possible that they remain somewhere in the active files at NASA. As of about 1980 the current files of the earth resources office contained almost nothing from before 1973, and no one in the office knew what might have happened to older subject files (chronological files are normally destroyed after three years). I did find copies of some useful older material in the office files of one member of the earth resources program office, Harry Mannheimer, then Landsat project manager.

Other offices in NASA concerned with Landsat tended to show the same pattern; I found little in the official record concerning the period I was studying but obtained useful material from individuals who had kept copies of key documents. The secretary at the Office of Applications, the division responsible for earth resources, said that that office did not keep old files on Landsat because that was the responsibility of the project office. Leonard Jaffe, who was responsible for the applications program at NASA from 1966 to 1978, allowed me to examine old chronological files, but these were limited almost entirely to letters signed by him. Richard Weinstein, then in the Technology Transfer Division of the Office of Applications, made available a file of documentation for the technology transfer program for Landsat in 1976 and 1977.

On a higher level the records were more systematically kept, and I found the largest amount of useful material in the retired subject files of the office of the administrator. These files provided a reasonably complete set of official correspondence between NASA and other government agencies, since such correspondence usually went from the administrator of NASA to the secretary of the agency or from the deputy administrator to the undersecretary. The administrator's files also contained memos on policy questions prepared by consultants and by the staff of the earth resources program office. Again, I also found useful material in active files; David Williamson, then assistant to the NASA administrator for special projects, allowed me to examine the minutes of the Earth Resources Survey Program Review Committee and the Interagency Coordinating Committee for Earth Resources Survey Program up to January 1975.

Finding material proved to be slightly easier at NASA's Goddard Space Flight Center. The current files of the Information Processing Division still contained a large amount of material on the development of the Landsat data processing system. The retired records at Goddard were also fairly complete; I used records of the information processing office, the ERTS [Landsat] study office (April 1967 to October 1970), and boxes of reports from principal investigators and contractors.

My luck was not as good at the Johnson Space Center. During my first visit to Johnson I was told that the records of the defunct Earth Resources Program Office had recently been destroyed, expect for presentation materials and budget statements. A selection of about twenty linear feet of documents on earth resources had been saved in the historical research collection at the Johnson Space Center History Office. Again, I also found useful material still in active files, particularly Leo Childs's files on the earth resources aircraft program and Bryan Erb's files on LACIE.

I did almost as thorough a survey of material on the Department of the Interior's EROS program as at NASA. At the EROS program office in Reston, Virginia, I found useful material in active files and in about twelve linear feet of old files that were stored in that office at that time (little material appeared to have been retired). I was particularly grateful for the use of two loose-leaf volumes of significant documents collected by William Fischer. At the EROS Data Center in Sioux Falls, South Dakota, I used thirteen boxes of subject files from the office of the chief of the center. People I interviewed also provided useful documents out of their own files.

I talked with people from other government offices and private industry, but I looked at only a small amount of unpublished material from those sources. Most of my material on the Department of Agriculture and the Corps of Engineers came from NASA files; because those agencies did not set up special offices for earth resources satellites, the relevant material is much more scattered. I surveyed what material was available at the National Archives from the Office of Management and Budget.

The resulting record is obviously spotty, but I saw a number of copies of most key documents, suggesting that I did not miss too many others. Writing a history of the space program poses special problems in finding, selecting, and interpreting sources. Government offices collect huge amounts of documentation, most of it so routine or so cautiously phrased as to be virtually useless. Each of the different types of sources—magazine and journal articles, official reports, briefing notes and internal reports, letters and memos, informal notes and records of telephone conversations, and interviews—gives a different type of picture. I found letters and memos to be the most useful sources because they were most likely to explain opinions on significant matters. Memos for the record and memos containing the minutes of meetings, in particular, provided more outspoken opinions than were available from any other written material.

Even in these cases, the source and standing of opinions expressed in letters and memos must be handled carefully. It is often not easy to tell whether a memo represents official policy or one person's opinion, and in letters it can be difficult to pin down whose opinion is involved. In the usual NASA practice for writing official letters, a letter was drafted by the person in the project office who knew the issue best. The letter was then sent up the hierarchy for concurrence by each level and then signed by the person in the hierarchy most appropriate to the recipient of the letter or the importance of the issue. In some cases this process could be traced by routing slips preserved with copies of letters. But in general it is important to remember that the person who signed a letter was not necessarily its author.

Space historians are particularly aware of these problems of finding and interpreting sources because those problems show up so clearly in the documentation. The incomplete documentation for Landsat was not entirely a disadvantage; Alex Roland tells of being faced with over 5200 cubic feet of unindexed documentation on the National Advisory Committee for Aeronautics. Historians always have to impose order on incomplete material; the key difference in writing recent history is that the participants' explanations of how they understood events are still available.

Guide to References to Manuscript Sources

WNRC and National Archives: For sources in Washington National Records Center and National Archives the information provided is folder title (often reflecting an agency coding system), box number, accession number, record group, and repository.

History Office, NASA Headquarters: The NASA History Office in Washington D. C. maintains a historical research collection of nonrecord copies of correspondence as well as many reports, articles, and internal documents. For these records the folder title is given.

History Office, Johnson Space Center: The Johnson Space Center in Houston, Texas, also maintains a historical research collection. At the time I used these documents they were organized chronologically by major subject and I used material from the earth resources section. The material has since been reorganized, and some of it may have been eliminated.

NASA Headquarters, Goddard Space Flight Center, and Johnson Space Center: Various offices and individuals at NASA Headquarters in Washington, D. C., the Johnson Space Center in Houston, Texas, and the Goddard Space Flight Center in Greenbelt, Maryland, allowed me access to their working files. The name of the individual or office is given, but folder titles are given only where files had been organized for office access. Some of the files of offices are record copies and may have been retired to the National Records Centers. The files of individuals would normally contain nonrecord copies. The record copy of the memos and correspondence I found in such files should have been retired to the National Records Centers by the appropriate office but often was not.

U. S. Geological Survey: Records consulted at the U. S. Geological Survey in Reston, Virginia, included the working files of the EROS Program Office and various individuals. Folder titles are given for the office files.

EROS Data Center: At the Department of the Interior's EROS Data Center in Sioux Falls, South Dakota, I used the working files of various individuals and the files of the Office of the Chief, which at the time were in storage at the center. Folder titles are given for the office files.

Environmental Research Institute of Michigan: Staff at this private company gave me access to reports and some correspondence files.

Published Sources

For a study such as this, secondary literature provides three kinds of context: broad issues about the process of technological change, specific themes in the history of the space program, and perspectives on the Landsat case study. The following list is intended to provide some introduction to the relevant literature, not to cover all material used in researching this study. See my review essay on space history in the July 1989 *Technology and Culture* for a systematic survey of the field.

The Process of Technological Change

Barnes, Barry, and David Edge, eds. *Science in Context: Readings in the Sociology of Science.* Cambridge: MIT Press, 1982.

Bijker, Wiebe E., Thomas P. Hughes, and Trevor Pinch, eds. *The Social Construction of Technological Systems: New Directions in the Sociology and History of Technology.* Cambridge: MIT Press, 1987.

Constant, Edward W., II. *The Origins of the Turbojet Revolution.* Baltimore: Johns Hopkins Univ. Press, 1980.

Edge, David O., and Michael J. Mulkay. *Astronomy Transformed: The Emergence of Radio Astronomy in Britain.* New York: John Wiley, 1976.

Gilfillian, S. C. *The Sociology of Invention.* Cambridge: MIT Press, 1963, first published 1935.

Hughes, Thomas P. "Inventors: The Problems They Choose, The Ideas They Have, and the Inventions They Make." In Patrick Kelly and Melvin Kranzberg, eds. *Technological Innovation.* San Francisco: San Francisco Press, 1978.

Hughes, Thomas P. "Technological Momentum in History: Hydrogenation in Germany 1900–1933." *Past and Present* (no. 44, 1969): 106–132.

Hughes, Thomas P. "The Development Phase of Technological Change." *Technology and Culture* 17 (1976): 423–431.

Hughes, Thomas P. *Networks of Power: Electrification in Western Society, 1880–1930.* Baltimore: Johns Hopkins Univ. Press, 1983.

Knorr-Cetina, Karen, and Michael J. Mulkay, eds. *Science Observed: Perspectives on the Social Studies of Science.* Beverley Hills: Sage, 1983.

Kuhn, Thomas. *The Structure of Scientific Innovation*, 2d edition. Chicago: Univ. of Chicago Press, 1970.

Latour, Bruno. *Science in Action.* Cambridge: Harvard Univ. Press, 1987.

Lauden, Rachel, ed. *The Nature of Technological Knowledge: Are Models of Scientific Change Relevant?* Dordrecht: Reidel, 1984.

MacKenzie, Donald, and Judy Wajcman, eds. *The Social Shaping of Technology: How the Refrigerator Got Its Hum.* Milton Keynes, Philadelphia: Open University Press, 1985.

Mansfield, Edwin. *The Economics of Technological Change.* New York: W. W. Norton, 1968.

Rogers, Everett M. *Diffusion of Innovations.* 1st and 3d editions, New York: The Free Press of Glencoe, 1962, 1983. Everett M. Rogers and F. Floyd Shoemaker, *Communication of Innovations: A Cross-Cultural Approach.* 2d edition. New York: The Free Press, 1971.

Schmookler, Jacob. *Invention and Economic Growth.* Cambridge: Harvard Univ. Press, 1966.

Usher, Abbot Payson. *A History of Mechanical Invention.* New York: McGraw-Hill, 1954.

Government Science and Technology

Brooks, Harvey. *The Government of Science.* Cambridge: MIT Press, 1968.

Dupree, A. Hunter. *Science in the Federal Government: A History of Policies and Activities to 1950.* Cambridge: Harvard Univ. Press, 1957.

Hough, Granville W. *Technology Diffusion: Federal Programs and Procedures.* Mt. Airy, Md.: Lombard Books, 1975.

Lambright, W. Henry. *Governing Science and Technology.* New York: Oxford Univ. Press, 1976.

Lambright, W. *Presidential Management of Science and Technology: The Johnson Presidency.* Austin: Univ. of Texas Press, 1985.

Roland, Alex. *Model Research: The National Advisory Committee for Aeronautics, 1915–1958.* Washington, D.C.: NASA SP-4103, 1985.

Sayles, Leonard R., and Margaret K. Chandler. *Managing Large Systems.* New York: Harper and Row, 1971.

Schubert, Glendon. *The Public Interest: A Critique of the Theory of a Political Concept.* Glencoe, Il.: The Free Press, 1960.

History of the Space Program

Bilstein, Roger E. *Stages to Saturn: A Technical History of the Apollo/Saturn Launch Vehicles.* Washington, D.C.: NASA SP-4206, 1980.

Burrows, William E. *Deep Black: Space Espionage and National Security.* New York: Random House: 1986.

Chapman, Richard LeRoy. *A Case Study of the U.S. Weather Satellite Program: The Interaction of Science and Politics.* Syracuse, N.Y.: Syracuse Univ. Press, 1967.

Davies, Merton E., and William R. Harris, *RAND's Role in the Evolution of Balloon and Satellite Observation Systems and Related U.S. Space Technology.* RAND Publication Series R-3692-RC. Santa Monica, Calif.: The RAND Corporation, Sept. 1988.

Doctors, Samuel I. *The NASA Technology Transfer Program: An Evaluation of the Dissemination System.* New York: Praeger Publishers, 1971.

Ezell, Edward C., and Linda Ezell. *The Partnership: A History of the Apollo-Soyuz Test Project.* Washington, D.C.: NASA SP-4209, 1978.

Ezell, Edward C., and Linda Ezell. *On Mars: Exploration of the Red Planet 1958–1978.* Washington, D.C.: NASA SP-4212, 1984.

Galloway, Jonathan F. *The Politics and Technology of Satellite Communications.* Lexington, Mass.: Lexington Books of D.C. Heath and Co., 1972.

Hacker, Barton C., and J. Grimwood. *On the Shoulders of Titans: A History of Project Gemini.* Washington, D.C.: NASA SP-4203, 1977.

Hall, R. Cargill. "Early U.S. Satellite Proposals." *Technology and Culture* 4 (1963): 410–434.

Hall, R. Cargill. *Lunar Impact: A History of Project Ranger.* Washington, D.C.: NASA SP-4210, 1977.

Hechler, Ken. *Towards the Endless Frontier: History of the Committee on Science and Technology, 1959–1979.* Washington, D.C.: U.S. House of Representatives, 1980.

Kenden, Anthony. "U.S. Reconnaissance Satellite Programmes." *Spaceflight* (July 1968): 243–262.

Kinsley, Michael E. *Outer Space and Inner Sanctums: Government, Business, and Satellite Communications.* New York: John Wiley, 1976.

Koppes, Clayton R. *JPL and the American Space Program: A History of the Jet Propulsion Laboratory.* New Haven: Yale Univ. Press, 1982.

Kvam, Roger A. "Comsat: The Inevitable Anomaly." In Stanford A. Lakoff, ed. *Knowledge and Power: Essays on Science and Government.* New York: The Free Press, 1966.

Levine, Arnold S. *Managing NASA in the Apollo Era.* Washington, D.C.: NASA SP-4102, 1982.

Logsdon, John M. "The Space Shuttle Program: A Policy Failure?" *Science* 232 (May 30, 1986): 1099–1105.

Logsdon, John M. *The Decision to Go to the Moon: Project Apollo and the National Interest.* Cambridge: MIT Press, 1970.

Mack, Pamela E. "Review Essay: Space History," *Technology and Culture* 30 (1989): 657–665.

Mark, Hans. *The Space Station: A Personal Journey.* Durham, N.C.: Duke University Press, 1987.

McDougall, Walter A. *. . . the Heavens and the Earth: A Political History of the Space Age.* New York: Basic Books, 1985.

Medaris, John B. *Countdown for Decision.* Princeton, N.J.: Van Nostrand, 1959.

Muenger, Elizabeth A. *Searching the Horizon: A History of Ames Research Center 1940–1976.* Washington, D.C.: NASA SP-4304, 1985.

Newell, Homer E. *Beyond the Atmosphere: Early Years of Space Science.* Washington, D.C.: NASA SP-4211, 1981.

Pierce, J. R. *The Beginnings of Satellite Communications.* San Francisco: San Francisco Press, 1968.

Roland, Alex. "The Shuttle: Triumph or Turkey?" *Discover* 6 (November 1985): 29–49.

Smith, Delbert D. *Communication Via Satellite: A Vision in Retrospect.* Leyden, Boston: A. W. Sijthoff, 1976.

Smith, Robert W. *The Space Telescopy: A Study of NASA, Science, Technology, and Politics.* Cambridge: Cambridge Univ. Press, 1989.

Smith, Robert W., and Joseph N. Tatarewicz. "Replacing a Technology: The Large Space Telescope and CCDs." *Proceedings of the IEEE* 73 (1985): 1221–1235.

Swenson, Lloyd S., Jr., J. Grimwood, and C. Alexander. *This New Ocean: A History of Project Mercury.* Washington, D.C.: NASA SP-4201, 1966.

U.S. Senate, Committee on Commerce, Science, and Transportation. *National Aeronautics and Space Act of 1958, As Amended, and Related Legislation,* 95th Congress, 2d session. Washington, D.C.: GPO, 1958.

Van Allen, James A. "Space Science, Space Technology, and the Space Station." *Scientific American* 254/1 (January 1986): 32–39.

Van Nimmen, Jane, Leonard C. Bruno, and Robert L. Rosholt. *NASA Historical Data Book, Volume 1: NASA Resources 1958–1968.* Washington, D.C.: NASA SP-4012, 1974, 1988. Linda Newman Ezell. *NASA Historical Data Book, Volume 2: Programs and Projects 1958–1968* and *Volume 3: Programs and Projects 1969–1978.* Washington, D.C.: NASA SP-4012, 1988.

Tomayko, James E. "NASA's Manned Spaceflight Computers." *Annals for the History of Computing* 7, (January 1985): 7–18.

Wells, Helen T., Susan H. Whitely, and Carrie E. Karegeannes. *Origins of NASA Names.* Washington, D.C.: NASA SP-4402, 1976.

White, Harold M., Jr. "International System and Space Law: An Introduction." In T. Stephen Cheston, Charles M. Chafer, and Sallie Birket Chafer. *Social Sciences and*

Space Exploration: New Directions for University Instruction. Washington, D.C.: NASA EP-192, 1984.

History of Remote Sensing and Earth Resources Satellites
Babington-Smith, Constance. *Air Spy: The Story of Photo Intelligence in World War II.* New York: Harpers, 1957.

Colwell, Robert N. *Monitoring Earth Resources from Aircraft and Spacecraft.* Washington, D.C.: NASA SP-275, 1971.

Compton, W. David, and Charles D. Benson *Living and Working in Space: A History of Skylab.* Washington, D.C.: NASA SP 4208, 1983.

Comptroller General of the United States. *Effects on Users of Commercializing Landsat and Weather Satellites.* Washington, D.C.: General Accounting Office, GAO/RCED-84-93, Feb. 24, 1984.

Darden, Lloyd. *The Earth in the Looking Glass.* New York: Anchor Press, 1974.

Dickson, Paul. *Out of This World: American Space Photography.* New York: Delacorte Press, 1977.

Freden, Stanley C., and Enrico P. Mercanti, eds. *Symposium on Significant Results Obtained from Earth Resources Technology Satellite-1, Volume 3: Discipline Summary Reports.* Greenbelt, Md.: Goddard Space Flight Center X-650-73-155, May 1973.

Harper, Dorothy. *Eye in the Sky: Introduction to Remote Sensing.* Montreal: Multiscience Publications Limited, 1976.

Heiman, Grover. *Aerial Photography: The Story of Aerial Mapping and Reconnaissance,* Air Force Academy Series. New York: Macmillan, 1972.

Holley, I. B. *Ideas and Weapons: Exploitation of the Aerial Weapons by the United States During World War I; A Study in the Relationship of Technological Advance, Military Doctrine, and the Development of Weapons.* Hamden, Conn.: Archon Books, 1971.

Joyner, Christopher C., and Douglas R. Miller. "Selling Satellites: The Commercialization of LANDSAT." *Harvard International Law Journal* 26 (Winter 1985).

Just, Theodore John. "The Origins of NASA Earth Resources Project (1963–1967)." Washington, D.C.: NASA HHN-96, 1969.

Katz, Amrom. "A Retrospective on Earth Resource Surveys: Arguments About Technology, Analysis, Politics, and Bureaucracy." *Photogrammetric Engineering and Remote Sensing* 42 (Feb. 1976): 189–199.

Lambright, W. Henry. "ERTS: Notes on a 'Leisurely' Technology." *Public Science* (Aug.-Sept. 1973): 1–8.

Lowman, Paul D., Jr., and Herbert A. Tiedemann. *Terrain Photography From Gemini Spacecraft: Final Geological Report.* Greenbelt, Md.: Goddard Space Flight Center #X-644-71-15, Jan. 1971.

Mack, Pamela E. "Space Science for Applications: The History of Landsat." In Paul A. Hanle and Von Del Chamberlain, eds. *Space Science Comes of Age: Perspectives on the History of the Space Sciences.* Washington, D.C.: National Air and Space Museum, 1981. Pp. 135–148.

Mack, Pamela E. "Satellites and Politics: Weather, Communications, and Earth Resources." In Alex Roland, ed. *A Spacefaring People: Perspectives on Early Space Flight.* Washington, D.C.: NASA SP-4405, 1985. Pp. 32–38.

Mack, Pamela E. "Commercialization, International Cooperation, and the Public Good." In Daniel S. Papp and John R. McIntyre, eds. *International Space Policy: Legal, Economic, and Strategic Options for the Twentieth Century and Beyond.* Westport, Conn.: Quorum Books, 1987. Pp. 195–202.

Office of Technology Assessment. *Civilian Space Policy and Applications.* Washington, D.C.: U.S. Congress, Office of Technology Assessment, OTA-STI-177, June 1982.

Office of Technology Assessment. *Remote Sensing and the Private Sector: Issues for Discussion—A Technical Memorandum.* Washington, D.C.: U.S. Congress, Office of Technology Assessment, OTA-TM-ISC-20, March 1984.

Office of Technology Assessment. *International Cooperation and Competition in Civilian Space Activities.* Washington, D.C.: U.S. Congress, Office of Technology Assessment, OTA-ISC-239, July 1985.

Proceedings of the First Symposium on Remote Sensing of Environment, 13, 14, 15 Feb. 1962. Ann Arbor, Mich.: Infrared Laboratory, Institute of Science and Technology, Univ. of Michigan, 1962.

Proceedings of the Second Symposium on Remote Sensing of Environment, 15, 16, 17, Oct. 1962. Ann Arbor: Infrared Laboratory, Institute of Science and Technology, Univ. of Michigan, 1962.

Quackenbush, Robert S., Jr. "Development of Photo Interpretation." In *Manual of Photographic Interpretation.* Virginia: American Society of Photogrammetry, 1960.

Rudd, Robert D. *Remote Sensing: A Better View.* North Scituate, Mass.: Duxbury Press of Wadsworth Publishing Co., 1974.

Nicholas M. Short, Paul D. Lowman, Jr., Stanley C. Freden, and William A. Finch, Jr., *Mission to Earth: Landsat Views the World.* Washington, D.C.: NASA SP-360, 1976.

Waldrop, M. Mitchell. "Imaging the Earth. II: the Politics of Landsat." *Science* 216 (Apr. 2, 1982): 40–41.

Zissis, Gerge J., et al. *Remote Sensing: A Partial Technology Assessment.* Ann Arbor: Environmental Research Institute of Michigan #123600-1, May 1977.

Interviews

NASA Headquarters, Washington, D.C.
Anthony Calio, June 12, 1978, June 29, 1978, April 5, 1979.
Leonard Jaffe, July 27, 1978
Bernard Nolan, June 25, 1979
William Raney, June 20, 1979
Donald Rogers, June 8, 1978, June 26, 1978
Richard Weinstein, August 28, 1980
David Williamson, August 27, 1979

Goddard Space Flight Center, Beltsville, Md.
Stanley Addess (Dept. of Interior), August 2, 1979
Stanley Freden, August 2, 1979
Thomas Ragland (Dept. of Interior), August 2, 1979
John Sos, July 25, 1979
William Webb, July 17, 1979

Johnson Space Center, Houston, Tex.
Ausley Carraway, September 11, 1979
Clifford E. Charlesworth, March 18, 1981
Leo Childs, September 10, 1979
Del Conte (Dept. of Agriculture), March 19, 1981
John E. Dornbach, September 11, 1979
Jack Eggleston, September 11, 1979
Bryan Erb, September 13, 1979
Bruce Jackson, September 12, 1979
Roy Hatch (Dept. of Agriculture), March 12, 1981
Joseph Loftus, September 11, 1979
Robert B. MacDonald, September 10, 1979
Barry H. Moore (Lockheed), March 20, 1981
Robert Piland, September 12, 1979
William E. Rice, September 10, 1979
Olav Smistad, September 12, 1979

Foreign Agricultural Service, Dept. of Agriculture, Houston, Tex.
Andrew Aaronson, March 18, 1981
James Hickman, March 18, 1981
Robert Spiers, March 19, 1981

Dept. of the Interior, Reston, Va.
Alden P. Colvocoresses, August 7, 1979
John DeNoyer, August 7, 1979
William A. Fischer, July 31, 1978
William Hemphill, July 31, 1978
Charles J. Robinove, July 31, 1978

EROS Data Center, Sioux Falls, S. Dak.
Ray Byrnes, May 4, 1981
Donald F. Carney, May 6, 1981
William C. Draeger, May 6, 1981
Donald T. Lauer, May 4, 1981
James R. McCord, May 7, 1981
R. A. Pohl, May 5, 1981
Fredericka A. Simon, May 6, 1981
Ralph J. Thompson, May 4, 1981
Allen H. Watkins, May 8, 1981

Environmental Research Institute of Michigan, Ann Arbor, Mich.
Marvin Holter, May 15, 1981
Fabian Polcyn, May 11, 1981
Donald Lowe, May 15, 1981
I. J. Sattinger, May 11, 1981
Gwynn Suits, May 11, 1981
George J. Zissis, May 13, 1981

Other Institutions
Peter Badgley, Arctic Intersciences Division, Office of Naval Research, Arlington, Va., August 7, 1978

John W. Jarman, U.S. Army Corps of Engineers, Washington, D.C., July 31, 1978

Archibald Park, General Electric, interviewed at NASA Headquarters, Washington, D.C., August 3, 1978

Robert Porter, Earth Satellite Corporation, Bethesda, Md., July 27, 1978

Willis Shapley, American Association for the Advancement of Science, Washington, D.C., August 23, 1979

Murray Strome, Canadian Centre for Remote Sensing, interviewed at the International Symposium on Remote Sensing of Environment, Ann Arbor, Mich., May 12, 1981

Index

AAP. *See* Apollo Applications Program
Active communication satellites, 23, 25
Administration. *See* NASA
Aerial photography, 31–33, 47, 48, 50, 75, 148, 166, 169, 174, 179, 185, 202
Aerial reconnaissance, 31–33, 52, 71, 89
Aerial survey, 50, 77, 88, 90, 123, 140, 145, 184, 191
Aerospace industry, 12, 25
Agency for International Development, 59, 184, 243n
Agricultural Adjustment Administration, 32
Agricultural survey. *See* Crops
Agriculture, 3, 5, 32, 37, 47, 48, 56, 57, 72, 73, 77, 113, 122, 127, 146–158, 169, 193, 245n
Agriculture, Department of, xiv, 11, 12, 14, 28, 51, 56, 57, 59, 64, 73, 75, 77, 84, 95, 111–113, 128, 132, 137, 146–158, 159, 163, 178, 185, 201, 203–204, 205, 210
AgRISTARS, 156
Airborne remote sensing, 48, 177. *See also* Remote sensing
Aircraft, 31–35, 38, 39, 40, 47–52, 71, 74, 77, 82–86, 88–90, 92, 137, 138, 140, 144, 145, 147, 161, 162, 165–166, 168, 169, 172, 177, 179, 184, 185, 218n, 219n. *See also* Earth Resources Aircraft Program
Air Force, 16, 34, 35, 40, 74, 91, 147, 216n, 219n
Alaska, 115, 229n
Altitude, 19, 20, 23–27, 33, 50, 51, 52, 89, 134, 147, 148, 172, 185, 228n

American Institute of Astronautics and Aeronautics, 89
American Institute of Mining Engineers, 144
American Society of Photogrammetry, 32
American Telephone and Telegraph (AT&T), 24–27
Ames Research Center, 168
Analog data processing, 117
Anders, William A., 109
Anderson, Clinton P., 162
Apollo, 9, 11, 17, 18, 46–47, 51, 52, 57, 68, 91, 92, 97, 98, 148, 152, 172, 218n, 219n, 220n
Apollo Applications Program (AAP), 52, 53, 57, 58, 66, 68, 82, 220n, 225n
Applications, xi, 5, 6, 8, 12–14, 15, 18, 22, 24, 28, 38, 51, 53, 57–58, 81, 96–97, 98, 104, 121–131, 142, 146–158, 163, 166–170, 174, 176, 177, 198, 201, 204, 209
Applications satellites, 15–30, 82, 89
Application system verification tests (ASVT), 165–167
Applications Technology Satellites, 82, 101
Archiving, 108, 109, 206, 229n
Argentina, 192
Arizona, 148
Army, 16, 49
Army Air Service, 32, 216n
Army Ballistic Missile Agency, 20
Army Corps of Engineers, 46, 57, 130, 210
Askew, Reuben, 169
Astronautics and Aeronautics, 89
Astronauts, 17, 31, 39, 40, 53, 184, 186

260   *Index*

Astronomers, 46, 217n
ASVT. *See* Application system verification tests
AT&T. *See* American Telephone and Telegraph
Atmosphere, 18, 35, 40, 50, 150, 215n
Augenstein, Bruno, 204
Australia, 108, 192
Automated satellites ("unmanned"), 47, 53, 54, 58, 219n

Babington–Smith, Constance, 32
Badgley, Peter C., 46–50, 52–55, 57, 58, 59, 62, 82, 95
Balloons, 23, 24, 31, 33, 34, 134
Battelle Columbus Laboratory, 167
Battlefield surveillance, 38, 71
Bell, Robert L., 137
Beltsville Space Center, 17. *See also* Goddard Space Flight Center
Bendix Corporation, 103, 104, 117, 242n
Benefits of satellite data, 6, 8, 13, 16, 20, 28, 39, 40, 57, 75, 76, 84, 89, 91, 92, 95, 111, 123, 126, 146, 167, 172, 175–176, 182, 186, 188, 198, 199–200, 206, 213n, 214n
Berkeley, University of California at, 50
Boeing Corporation, 101
Bombing, 32, 33
Brazil, 184, 185, 189, 192, 194, 243n
Budgetary considerations, xiii, 4, 5, 10, 11, 12, 18, 46, 66, 80–93, 97, 118, 133–137, 154, 197, 199, 200, 202, 203, 205, 220n, 226n
Budget cuts, 10, 75, 84, 88, 90, 97, 118, 135, 154, 200, 202, 203, 220n
Budget process, xiii, 4, 5, 66, 80–81, 136
Bureau of the Budget, xii, 57, 61, 75, 76, 80–93, 103, 107–111, 118, 134, 150, 151, 172, 197, 200, 209, 225n, 228n. *See also* Office of Management and Budget
Butz, Earl, 151
Byerly, Theodore, 150

California, 17, 41, 148, 149, 168, 169
California, University of, 50, 148
Cameras, 19, 32–35, 37, 40, 42, 48, 51, 52, 53, 54, 59, 61, 67, 68, 71, 75, 78, 162, 193, 194, 215n, 216n, 220n, 227n. *See also* Photographic sensors and Television cameras
Canada, 154, 155, 166, 191–192
Carlson, W. Bernard, xiv
Carter, Jimmy, 6, 10, 29, 181, 196, 205
Cartography, 49, 57, 59, 99, 109, 142, 206
CCD. *See* Charge-coupled device
Central America, 188
Central Intelligence Agency (CIA), 35, 88, 95, 96, 131
Charge-coupled device, 36
Charlesworth, Clifford E., 128
Chevron Oil Corporation, 174, 175
Childs, Leo, 50–51
China, People's Republic of, 154
Christenson, Albert T., 165
Civilian satellites, 3, 6, 16, 31, 35–39, 53, 91, 181, 185, 187, 193, 204
Clark, John, 98
Classified information, 4, 11, 33–42, 50, 60, 61, 72, 75, 90, 91, 131, 179, 227n
Climate. *See* Weather
Clouds, 19, 20, 129, 218n
Coarse resolution. *See* Resolution, coarse
Coast Guard, 165, 166
Coastlines, 111, 129, 131, 175
Color enhancement, 177, 179
Color photographs, 40, 70–72, 78, 108, 112–115
Colvocoresses, Alden P., 49, 99, 109, 142
Colwell, Robert N., 49, 148
Commerce, Department of, 20, 22, 206
Commercialization, 3, 24–25, 27, 29, 136, 148, 171–182, 193–195, 206, 208, 210, 247n
Communication, 6, 8, 39, 136, 165, 166, 211
Communications Satellite Corporation (ComSat), 25, 27, 171
Communications satellites, 3, 15, 18–20, 23–27, 31, 82, 171, 199
Competition, bureaucratic, 4, 9, 11, 136, 195, 208–211
  from foreign sources 195, 204, 210 (*see also* Satellites, foreign)
  government vs. private industry, 126–127, 138, 144, 171–182
Computer analysis, 27, 116, 127, 129, 139, 143, 144, 148, 149, 153, 177

Computers, 27, 103, 104, 108, 109, 114, 117, 127, 129, 130, 140, 142, 143, 148, 149, 152, 153, 157, 164, 177, 179, 232n
Computer tapes, 113–117, 139, 140, 152, 179, 215n, 242n
ComSat. *See* Communications Satellite Corporation
Conferences, 38–39, 53, 89, 127, 168, 174, 185, 186, 188
Conflict, xi, 4, 5, 6, 9, 11, 13–14, 15–19, 22–25, 31, 42, 45, 53, 55, 56–65, 66, 69, 74, 76, 80–93, 100, 133, 136, 151–152, 168, 171–172, 195, 196–207, 208–210
Congress, xii, xiii, 10, 11, 12, 16, 18, 23, 25, 52, 57, 59, 65, 80, 81, 84, 85, 86, 130, 151, 159, 162, 182, 196, 201, 206, 220n, 226n
Conoco (Continental Oil Company), 174
Constituency of Landsat, 52, 96, 122, 131, 132, 159, 201, 202, 209, 219n
Construction of Landsat system, 85, 86, 94, 96, 100–106, 117, 135, 137
Contractors, 94, 100–106, 113, 130, 179, 180
Control, xi, 5, 9, 11, 16, 25, 29, 61, 62, 64, 73, 79, 97, 99, 100, 132, 136, 137, 168, 169, 172, 193, 196–207, 209
Cooper, L. Gordon, Jr., 39–40
Cooperation, interagency, 11, 94, 98–100, 168, 175
international, 10, 183–195, 243n
Corn Blight Watch, 52, 147, 148, 150
Cornell University, 199
Correction of data, 103, 104, 110, 112, 113, 114, 179, 191, 232n
Cosmos, 193
Cost issues. *See* Economic considerations
Cost–benefit analysis, 80, 91–93, 146, 150, 156, 164, 172, 175,199, 201, 203, 205, 228n, 238n
Cousteau, Jacques, 130
Coverage of satellite data, 33, 39, 61, 78, 89, 148, 176, 210, 229n
Crabill, E.D., 84
Crop Condition Assessment System, 156, 157
Crops, 3, 28, 47, 48, 72, 73, 77, 109, 122, 127, 146–158

Crop yield, measurement of, 28, 49, 127, 146–158, 177, 178, 238n
Differentiation, 32–33, 47–48, 72, 148, 149, 153, 154–155
Disease, 3, 32, 52, 73, 146, 147
Curtis, Carl, 135

Data analysis, 50, 95, 132–145, 147–148, 151, 152, 155, 171–182, 185, 194, 202
Data center (EROS), 85, 86, 113, 115, 117, 128, 132–145, 157, 160, 167, 175–179, 181, 189, 209, 242n
Data processing, 7, 27, 66, 70, 76, 79, 85, 86, 93, 95, 98, 102–106, 107–118, 127, 132–145, 171–182, 191, 197, 229n, 232n, 245n
Data use, 121–131, 132–145, 146–158, 159–170, 171–182, 183–205, 209
Data distribution, 85, 112, 123, 125–126, 132–137, 139–140, 145, 159–170, 171–182, 183–195, 196
Declassification, 11, 36, 38, 39, 58, 72, 74, 223n
Defense, Department of, 4–5, 11, 15–17, 22, 24, 27, 31, 33–37, 39, 48, 49, 56, 58, 65, 69, 71, 72, 74, 75, 82, 90, 91, 130–131, 228n, 232n
Definition of new technology, xi, 4, 13, 30, 56, 66, 89–90, 146, 152, 181, 183, 195, 197, 198, 207, 209–211, 233n, 247n
Demand for data, 171–182
DeNoyer, John M., 95, 111
Design of data processing system, 85, 104, 107–118, 132–145
Design of Landsat, 4, 5, 14, 66–79, 85, 88, 93, 94–106, 107–118, 121–122, 126, 132, 157–158, 169, 197–198, 202, 208, 210
Design phase of technology, 4, 19. *See also* Development phase
Developing countries, 47, 59, 89, 181, 183, 186, 188, 189, 193, 194, 206
Development phase of technological change, xi, 3–7, 18, 19, 39, 45, 60–64, 66–79, 93, 102, 107, 121–123, 145, 164, 165, 192, 197–199, 207–209
Diffusion of technology, 7, 8, 121–131, 132, 162, 163, 208, 209, 233n. *See also* Dissemination
Digital analysis of data, 117, 129, 145, 167, 178

Digital data processing, 103–106, 108, 113, 116, 117, 129, 143, 145, 157, 167, 177, 178, 200
Discoverer, 35
Discoveries, Landsat, 122, 127, 131
Disease, 32, 146
Dissemination (distribution) of data, 85, 108, 109, 112, 113, 115, 121–131, 132–145, 163, 171–182, 183–195, 196
Doctors, Samuel I., 163
Draeger, William, 144
Drought, 32, 109

Early Bird, 27
Earth observation, 46, 65, 74, 88, 220n
Earth Observations Office (NASA), 128
Earth Observations Satellite, 78
Earth Observatory Satellite, 198, 200
Earth Observing Satellite Corporation (Eosat), 206
Earth resources, xii, 4, 40, 42, 45–47, 50, 51–55, 56–59, 66–70, 73, 82, 83, 87, 89, 91, 97, 98, 100, 108, 123, 125, 127, 133, 146–158, 166, 174, 177, 183–195, 196–207
Earth Resources Aircraft Program, 50–52, 67, 69, 85, 86, 90, 97, 113, 161, 184, 219n
Earth Resources Laboratory, 161, 166, 167, 168
Earth Resources Observation Satellites/System (EROS), 59, 61–65, 67–69, 74, 83–86, 91, 101, 112, 115, 117, 125, 128, 132–145, 157, 160, 161, 167, 204, 208
Earth Resources Observation System (EROS) Data Center, 85–86, 112–113, 115, 117, 128, 132–145, 157, 160, 167, 175–179, 181, 189, 209, 242n. *See also* Data center, EROS
Earth Resources Observation System Program Office, 99, 142
Earth resources observatories, 54, 55, 59
Earth Resources Program Office (NASA, Johnson Space Center), 94, 95, 109, 128, 130, 151, 177
Earth resources satellites, xii, 9, 10, 13, 15, 19, 27–30, 31–42, 49, 52–55, 56–65, 69, 78, 82–93, 94, 101, 111, 117, 122, 123, 125, 127, 132, 135, 146, 148, 150, 170, 171–182, 183–195, 196–207, 208–211
Earth Resources Survey Program (ERS), 85, 91, 97, 99, 117, 127, 187, 220n
Earth Resources Survey Program Review Committee (ERSPRC), 73, 98, 99, 111
Earth Resources Techniques Development Laboratory, 174
Earth Resources Technology Satellite (ERTS, renamed Landsat), 4, 70, 84, 90, 92, 93, 98, 99, 111, 116, 143, 164, 173, 197, 201, 226n
Earthsat. *See* Earth Satellite Corporation
Earth Satellite Corporation (Earthsat), 92, 95, 176, 178
Echo, 23, 24,
Economic considerations, 20, 22, 29, 61, 74, 76, 86, 103, 110, 114, 131, 134, 146, 150, 160, 179, 186, 192, 200, 201, 205, 209
Eisenhower, Dwight David, 11, 16, 34,
Electron Beam Recorder, 68, 69, 114, 116, 192
Engineers, 3, 8, 18, 24, 26, 32, 34, 45, 48, 50, 66–69, 73, 77, 79, 96, 97–99, 101, 104, 107, 109, 110, 116, 117, 118, 136, 152, 198, 199, 202, 229n
Environment, 89, 177, 185, 193
Environmental Research Institute of Michigan (ERIM), 37–38, 48–49, 71, 92, 130, 148, 177
Environmental Science Services Administration, 21
Eosat. *See* Earth Observing Satellite Corporation
ERIM. *See* Environmental Research Institute of Michigan
EROS. *See* Earth Resources Observation Satellites
ERS. *See* Earth Resources Survey Program
ERSPRC. *See* Earth Resources Survey Program Review Committee
ERTS. *See* Earth Resources Technology Satellite or Landsat
Experimental vs. operational programs, xii, 11, 13, 19, 22, 23, 28, 29, 60, 62, 64, 65, 80, 92, 93, 96, 100, 103, 108, 110, 116, 118, 122, 125, 132, 133, 136, 145, 150, 156, 164,

169, 171, 186, 187, 196–198, 208–209
Extractive industry, 144, 172, 175, 210. *See also* Mining

Fear of failure, 107, 108, 110
Faults, geological, 48, 123, 127, 172, 173
Feyerherm, Arlin, 153
Film, photographic, 27, 43, 35, 36, 53, 68, 71, 74, 103, 114, 116, 215n. *See also* Photographic sensors
Film return satellites, 35, 74–75, 89,
Fischer, William, 58–61, 63, 90, 101, 133
Fletcher, James C., 92, 100, 162, 164, 169, 174
Flooding, 28, 32, 128, 130, 161, 206
Forecasting, crop, 28, 47, 49, 146–158, 177, 178, 238n. *See also* Crops
weather 20, 110, 213n. *See also* Weather and Weather satellites
Foreign Agricultural Service, 156, 157, 159, 210
Foreign governments, 3, 123, 144, 146, 151, 175, 177, 180, 183–195, 210
Foreign users, 3, 123, 140, 141, 143–145, 175, 177, 180, 182, 183–195, 203, 210
Forestry, 49, 50, 113, 142, 161, 162
France, xii, 3, 29, 187, 193–195, 204, 210
Freden, Stanley C., 128
Frutkin, Arnold W., 180, 187
Funding, 5, 8, 12, 14, 16, 19, 24, 28, 34, 38, 46, 57–59, 64, 67, 80–93, 94, 95, 100–104, 108, 110, 112, 115, 117, 126, 128, 132–137, 150, 151, 156, 160, 161, 164, 168, 169, 171, 188, 189, 199–203, 220n

Gemini, 40, 41, 52, 172, 184
General Electric, 101–104, 116–117, 179, 180, 191, 199
General Accounting Office (GAO), 113, 155, 168
Geographers, 3, 37
Geography, 59, 71, 112–113, 126, 186
Geological Survey. *See* United States Geological Survey
Geologists, 37, 40, 46, 49, 72, 95, 127, 218n
planetary, 3, 46, 218n

Geology, 3, 5, 28, 32, 48, 49, 56, 57, 59, 73, 112, 123, 126, 139, 172–176, 206
Geometric distortion and correction of data, 69, 112–114, 179, 191, 232n
Geophysics, 67, 96
Geostationary orbits, 23
Geosynchronous satellites, 20, 23, 25, 27
Gilfillan, S. C., 7
Goddard Space Flight Center, 17, 18, 66–67, 69, 70, 73, 77, 96–104, 109–117, 124, 125, 127, 128, 137, 168, 179, 191, 229n
Government context, problems of, xi, xii, 3–14, 171–182
Government users of Landsat data, federal, 3, 28 123, 125, 141, 144, 167, 171, 178, 182, 208, 210
foreign, 3, 25, 123, 140, 141, 144, 178, 183–195, 210
local, 12, 28, 123, 141, 142, 144, 159–170
state, 12, 28, 29, 123, 139, 141, 144, 159–170, 178, 182, 205, 206, 208–210
Ground stations, 24, 27, 113, 165, 183, 189–192, 194, 229n
Ground surveys, 49, 92, 142, 147, 150, 172, 184, 193
Ground truth, 49, 142, 153, 154, 184, 193

Hardin, Clifford, 147
Hardware, 4, 6, 8, 9, 16, 22, 32, 140, 163, 176, 179, 233n
Heat sensors, 33. *See also* Thermal sensors
Hickel, Walter J., 92, 135
History of Landsat, xii, 4, 6, 7, 11, 14, 15, 27–30, 54, 66–79, 84
History of technology, xii, 3, 4, 6, 7, 88, 210, 211
Holter, Marvin R., 71
House Subcommittee on Space Science and Applications, 10, 85
Houston (Texas), 17, 50, 52, 64, 97, 161, 174
Hughes, Thomas P., xiv, 7
Hughes Aircraft Corporation, 25, 26, 72, 104, 206, 223n
Hurricanes, 20
Hydrologists, 3, 58, 122, 127, 129–131
Hydrology, 57, 59, 126, 129–131

IBM Corporation, 103, 117
ICCERSP. *See* Interagency Coordinating Committee for Earth Resources Survey Program
Ice, 115, 128, 165, 166, 191
ICEWARN, 165, 166
Image data, 4, 19–20, 27, 36, 37, 49, 71, 76, 103, 110, 113–114, 117, 124, 131, 137–139, 141, 143, 144, 145, 149, 156, 157, 162, 169, 173–179, 186, 189, 191, 192, 215n, 217n, 220n, 229n, 232n
Image Processing Division (Goddard Space Flight Center), 113, 116
Indexing system, 138
India, 131, 154, 192, 194
Industrial use of Landsat data. *See* Private industry
Information extraction, 27, 32, 108–110, 123, 125, 129, 164, 174, 176
Information systems, 142, 143, 157, 169
Infrared, 32, 33, 40, 51, 71, 72–73, 147, 191, 200, 204, 245n
  far–infrared band, 32, 51, 71, 77, 78, 200, 245n
  film, 32, 33, 48,
  photography, 48, 50
  scanning, 35, 54, 55, 74,
Infrared Laboratory, 37, 38, 71. *See also* Environmental Research Institute of Michigan
Innovation, xi, 3, 4, 7, 8, 12, 14, 37, 95, 121, 163, 171, 179, 197, 212n, 213n, 232n, 233n
Insects, 32, 185
Institute of Science and Technology (University of Michigan), 37–38, 177. *See also* Environmental Research Institute of Michigan and Willow Run Laboratory
Instruments, 39, 40, 50, 54, 55, 66, 67, 98, 126, 203. *See* Cameras, Scanners, and Sensors
Interagency Coordinating Committee for Earth Resources Survey Program, 99, 117
InterAmerican Development Bank, 144
Intergovernmental Science, Engineering, and Technology Advisory Panel (NASA), 169

Interior, Department of, xiv, 11, 28, 41, 46, 51, 56, 57–65, 67, 69, 73, 75–77, 82–87, 91, 92, 95, 96, 99, 101, 111–113, 125, 127, 128, 132–145, 150, 160, 161, 167, 178, 184, 199–205, 208, 209, 211, 243n
International use of Landsat data, 140–145, 183–195
International Geophysical Year, 17, 18
International law, 37, 183, 185–188
International relations, 24, 62, 180, 181, 183–195, 243n
Interpretation of data, 47, 49, 51, 52, 125, 130, 175, 193
Invention, 6, 7, 121, 156, 163, 167, 175
Italy, 192

Jaffe, Leonard, 54, 58–60, 62–64, 75, 89, 98, 104, 204
Jamison, David W., 160
Japan, 192, 194, 195, 204, 210
Jet Propulsion Laboratory (JPL), 117
Johnson, Lyndon B., 62, 84
Johnson Space Center (Houston), xiii, 17, 45, 50, 51, 52, 64, 66, 67, 69, 77, 97, 98, 128, 130, 147, 150, 151, 160–162, 166, 168, 174, 179, 200, 219n
Joint research, 29, 130, 168, 174, 175, 185, 193, 200, 206
JPL. *See* Jet Propulsion Laboratory

Kansas, 153
Kansas State University, 153
Kansas, University of, 48
Karth, Joseph E., 10, 85–87, 92
Katz, Amrom H., 88–89, 90, 91, 227n, 228n
Kennedy, John F., 11, 17, 18
Kraft, Christopher C., Jr., 162, 168

Laboratory for Application of Remote Sensing (Purdue University), 148
Laboratory for Atmospheric and Biological Sciences (Goddard Space Flight Center), 96
LACIE. *See* Large Area Crop Inventory Experiment
Lambright, W. Henry, 10, 19
Land Remote Sensing Commercialization Act, 206
Land use, 5, 28, 32, 49, 135, 142, 143, 146–158, 177, 179, 182

Langley Research Center, 24
Large Area Crop Inventory Experiment (LACIE), 146–158, 163, 165, 209
Laser, 54, 68
Latin America, 59, 185
Latour, Bruno, 5, 233n
Lattman, L. H., 38
Lauer, Donald, 145
Launches, xii, 3, 4, 5, 10, 18, 19, 20, 23–27, 29, 34–36, 51, 52, 59, 61, 70, 77, 87–89, 94, 96, 97, 99, 101, 104, 107, 115, 116, 122, 125, 127, 136, 139, 154, 171, 172, 174, 178, 180, 185, 186, 189, 191–195, 196, 198–200, 210, 220n, 233n
Lockheed Corporation, 34, 102, 154
Logsdon, John M., xiv, 9
Louisiana, 161
Low-altitude satellites, 19, 23–27
Low, George M., 160, 175
Luce, Charles F., 62, 63, 64
Lunar exploration, 17, 18, 46, 47, 51, 218n. *See also* Moon
Lunar Orbiter, 35, 46, 101, 110

McCabe, Francis X., 147
MacDonald, Robert B., 150, 151
McDougall, Walter, 34
Magnetometer, 54, 175
Magnetic sensors, 191
Mali, 189
Management of Landsat, 94–98
Manned Spacecraft Center, 17, 47. *See also* Johnson Space Center
Manned spaceflight program, 46. *See also* People in space
Manned Space Science, 184
Mapping, 31, 32, 51, 74, 75, 114, 122, 128, 130, 131, 142, 173, 176, 183, 189, 218n
Maps, 20, 21, 28, 32, 69, 136, 139, 142, 173, 175, 177, 186, 189, 194, 218n
Mark, Herman, 165–166
Martin Marietta Corporation, 174
Maryland, 17, 113, 169
Mathews, Charles W., 152
Mathias, Charles, Jr., 169
Mayo, Robert P., 92
Media, 18
Mercury, 39, 40, 42, 52, 184
Meteor, 193
Meteorologists, 20, 21,

Meteorology, 19, 42, 67, 113, 133, 146, 171, 186, 206. *See* Weather satellites
Mexico, 184, 185, 187
Michigan, 38, 49, 177
Michigan, University of, 37–38, 47, 48, 51, 71, 148, 177
Michigan Aeronautical Research Center, 71. *See also* Environmental Research Institute of Michigan
Microwave sensors, 53, 54, 198
Military, 15–18, 22, 31, 33–39, 42, 51, 61, 66, 71, 74, 88, 131, 134, 186, 193, 217n, 223n
Military satellites, 31, 131, 193. *See also* Reconnaissance satellites
Minerals, 28, 48, 144, 173, 175, 176, 243n
Mining, 144, 172–175, 184, 191, 210
Missile programs, 17, 71
Missions Utilization Office (Goddard Space Flight Center), 128
Mississippi, 161, 166, 167
Mississippi Test Facility (NASA), 161, 166
Monitoring, 18, 20, 27, 28, 37, 42, 45, 108, 129, 142, 144, 148, 154, 157, 166, 176, 177, 191
Moon, 17, 18, 35, 46, 47, 52, 68, 117, 218n. *See also* Lunar exploration
Morley, Lawrence, 191
MSS. *See* Multispectral Scanner
Multilinear array sensors, 194
Multispectral Scanner (MSS), 35–36, 51, 53, 54, 70–73, 77, 78, 91, 114, 115, 147, 160, 194, 200, 202, 203, 223n
Multistage sampling, 51
Mundt, Karl E., 134, 135, 136

NACA. *See* National Advisory Committee for Aeronautics
NASA. *See* National Air and Space Administration
National Academy of Engineering, 164
National Academy of Science, 89, 92
National Advisory Committee for Aeronautics (NACA), 17
National Aeronautics and Space Act, 16, 18, 74
National Aeronautics and Space Council, 109

266   Index

National Air and Space Administration (NASA), xi, xii, xiii, xiv, 3–14, 15–30, 31, 33–36, 39–42, 45–55, 56–65, 66–93, 94–106, 107–118, 121–131, 146–158, 159–170, 171–182, 183–195, 196–207, 208–211, 219n
  Administration, xiii, 5, 10, 12, 14, 17, 18, 45, 56, 61, 62, 66, 72–74, 76, 80, 81, 91–93, 94–100, 107, 110, 117, 122, 123, 125, 128, 129, 131, 136, 146, 150–152, 154, 159, 160, 162, 164, 166, 168, 171–175, 178, 180, 184, 192, 196–207, 209
  Constituency, 7, 12, 14, 17, 18, 52, 122, 131, 159, 209
  External conflicts, 13–14, 22–23, 31, 56–65, 133, 151, 168, 200, 208–211
  Headquarters, xiii, 18, 46, 52, 95, 96–98, 99, 109, 111, 112, 126, 128, 137, 151, 162, 165, 166, 168, 181, 184, 229n
  History, xiii, 3, 6, 9, 14, 15–30, 34, 81,
  Internal Conflicts, 17–18, 168, 200, 208–211
  International Participation Program, 184
  Office of Applications, xiii, 104, 165, 204
  Structure, 12–14
  Technology Utilization Program, 163
National Air and Space Museum, xiii
National Oceanic and Atmospheric Administration (NOAA), 6, 21, 29, 57, 112, 113, 128, 137, 146, 151, 153, 156, 203, 205
National prestige. *See* Prestige, national
National Research Council Committee on Remote–Sensing Programs for Earth Resources Surveys, 201–202
National Research Council Committee on Space Program for Earth Observations, 73–74
National Science Foundation, 12, 177
National security, 4, 36, 74–76, 88, 91
National Security Council, 58
National Space Technology Laboratory (formerly Mississippi Test Facility), 161, 166, 167
Natural resources, 27, 37, 46, 47, 61, 62, 89, 160, 176, 182, 184, 186, 187, 189, 193, 200, 205, 220n

Naugle, John E., 86, 87
Naval Oceanographic Office, 57, 59, 185
Naval Research Laboratory, 18
Navy, 34
Nebraska, 135, 169
Newell, Homer E., Jr., 54, 64, 81, 82, 84, 86,
New Mexico, 162
Nimbus, 20–22, 47, 55, 73, 82, 101, 102
Nixon, Richard M., 10, 81, 82, 87, 88, 91, 134, 135, 147, 180, 186, 225n
NOAA. *See* National Oceanic and Atmospheric Administration
Nordberg, William, 73, 96, 98, 99, 111, 112
North Dakota, 155
Norwood, Virginia, 72

Oceanography, 47, 57, 113
Office of Management and Budget (OMB), 10, 11, 13, 29, 81, 87, 100, 104–105, 151, 154, 156, 157, 164, 199–203, 225n. *See also* Bureau of the Budget
Office of Manned Space Science, 46, 47, 52, 53. *See also* Office of Space Science and Applications
Office of Naval Research, Geography Branch, 38
Office of Science and Technology, 100, 150
Office of Science and Technology Policy, 204
Office of Space Science and Applications (NASA), 46, 54, 55, 81, 86, 98, 145, 152, 165, 204
Ohio, 17
Ohio State University, 48
Oil exploration, 3, 28, 48, 123, 126, 127, 137, 144, 172, 174–176
OMB. *See* Office of Management and Budget
Open skies issue, 37, 154, 180, 185–187
Operating agencies, 57. *See also* Interior, Department of, and National Oceanic and Atmospheric Administration
Operational (vs. experimental) progams, xii, 5, 6, 8, 11, 13, 19, 22, 23, 25, 27, 29, 30, 53, 57, 58, 62–63, 80, 92, 93, 99, 103, 110, 111, 116,

118, 121–131, 132–145, 146, 148, 152–158, 164, 170, 171, 176, 181, 186, 187, 188, 196–207, 208–211
Orbiting Astronomical Observatory, 55
Orbiting Geophysical Observatory, 96, 102
Oregon, 169, 177
Oregon State University, 177
Organization of American States, 63
Overflights, 33, 154. *See also* Open skies issue and U–2

Padrick, Ben D., 168, 169
Paine, Thomas O., 87
Park, Archibald B., 73, 75, 95, 111, 147
Parker, Dana, 38
Passive satellites, 23, 24
Peaceful purposes, 16, 18, 184, 186
Pecora, William D., 60, 61, 63, 64, 86, 87, 98
Pennsylvania State University, 38, 177
Pentagon. *See* Defense, Department of
People in space ("manned" spaceflight), 16, 17, 18, 40, 45–47, 52, 53, 55, 58, 82, 83, 97, 219n
Petroleum. *See* Oil exploration
Philco–Ford Corporation, 101
Photogrammetric Engineering, 32
Photogrammetry, 31, 37, 160, 189
Photographic sensors, 48, 59, 73–75
Photographs, 21, 31–33, 35, 38, 40–42, 49, 102, 103, 109, 112–117, 123, 136–141, 145, 150, 172, 174, 179, 184–186, 193, 202, 215n
Photography, 31–33, 34–35, 37, 39, 40, 48, 51, 54, 55, 67, 68, 74, 75, 89, 91, 148, 166, 220n
Photo–interpreters, 32, 109, 116, 129, 144, 145, 153, 154, 156, 167, 174
Physicists, 46
Piland, Robert O., 166
Planetary exploration, 46, 66, 117
Planning Research Corporation, 199
Pohl, Rus, 134, 135
Policy, xii, xiii, 9, 10, 58, 61, 64, 81, 87, 94–99, 126, 127, 137, 150, 157, 163–164, 180, 181, 196, 198, 204, 225n, 229n
Politics, 5, 6, 9–12, 16, 18, 23, 25, 27, 52, 53, 56–57, 60–61, 66–67, 81, 84, 89, 91, 101, 108, 122, 133–135, 142, 145, 150–151, 154, 158, 159, 164, 169, 183–195, 196– 197, 212n
Politics, international, 183–195
Porter, J. Robert, 95, 178, 179
Precision–processed images, 103, 112, 114, 117, 191
Presidential decision–making, xii, xiii, 9, 10, 15, 25, 28, 80–82, 85, 86, 87, 88, 91, 92, 100, 109, 133, 135, 150, 181, 196, 204, 205, 210, 219n, 220n, 226n
President's Science Advisory Committee, 16, 100, 150
Press, role of. *See* Media
Prestige, national, 15, 16, 17, 218n
Prices charged for data, 136, 100, 205
Principal investigators (Landsat), 112, 115, 126, 127–129, 137, 141, 160, 172, 175, 189
Private industrial use of Landsat data, 29, 95, 171–182
Private industry, xi, xii, 3, 6, 10, 14, 24, 25, 27–29, 74, 94–96, 101, 104, 123, 134, 138, 139, 141, 143, 144, 164,171–182, 196, 204–207, 208–210, 214n, 242n, 243n
Privatization, xii, 6, 23, 29, 171–182, 188, 196–207. *See also* Commercialization
Procurement process, 199
Production of Landsat data, 181
Profit, 6, 13, 23, 24, 25, 29, 146, 150, 170, 172, 177, 178, 181, 182, 206
Project Michigan, 38, 71
Project RAND. *See* RAND Corporation
Proposals for experiments with Landsat data, 112, 126, 189
Proprietary data, 172, 174, 180, 181
Publicity, 14, 29, 130, 134, 139, 148, 162, 165, 208, 209
Public relations, 184
Public service, 6, 113, 126, 127, 137, 144, 171, 182, 197, 206, 210, 213n, 214n, 247n
Purdue University, 47, 147, 148, 150, 177, 243n
Pushbroom scanner, 204

Questar telescope, 40

Radar sensors, 33, 48, 51, 54, 165, 218n, 219n

Radio, 27, 35, 42, 68, 69, 113, 134, 217n, 229n
Radiometers, 54
Radio waves, 23, 37, 113
Ragland, Thomas M., 96
RAND Corporation, 34, 89, 91, 216n, 227n, 228n
Ranger, 68, 69
RCA, 25, 59, 68, 69, 73, 101, 102, 133, 135, 206
Reagan, Ronald, 206
Real–time imagery, 36
Reconnaissance satellites, 4, 5, 11, 16, 18, 20, 31, 33–39, 58, 65, 88, 90, 91, 95, 102, 179, 185, 187, 193, 215n, 216n, 227n, 228n
Regional Dissemination Centers (NASA), 163
Regional use of Landsat data, 159–170, 208, 210
Reifel, Ben, 134
Relay, 23, 25
Remote sensing, 6, 37–40, 42, 46–53, 57–58, 73, 82, 125, 142–144, 147, 148, 156, 160, 161, 165, 166, 169, 174, 177, 181, 183–185, 187, 189, 191, 194, 204, 205, 223n, 243n
Repetitive coverage, 27, 39, 42, 49, 61, 72, 76, 89, 91, 115, 148, 176, 201
Research and development, xi, 5–9, 12–14, 15–30, 32, 53, 61–65, 66, 107, 121–123, 152, 156, 157, 165, 172, 197, 198, 200, 206, 208
Resolution, 36, 40, 47, 54, 58, 68, 69, 74, 75, 78, 89, 90, 91, 92, 99, 103, 115, 193, 194, 217n
 coarse, 4, 27, 35, 36, 40, 75, 76, 90, 91, 148, 154, 186, 193
 moderate, 186, 194
 fine (high), 33, 36, 40, 50, 53, 59, 68, 70, 72, 74, 76, 78, 88, 89, 91, 177, 186, 187, 189, 198, 204
Resource management, 3, 5, 13, 14, 27, 28, 37, 40, 50, 62, 112, 159, 160, 176, 182, 200, 205, 210
Resource managers, 29, 42, 112, 115, 128, 131, 142, 143, 150, 160, 164, 165, 208, 209, 219n
Return beam vidicon, 53, 68–73, 78, 114, 133
Rice, Donald B., 100
Robinove, Charles J., 58, 59, 60, 91, 101
Rockets, 16, 34, 166

Rogers, Everett M., 167
Roland, Alex, xiv

Sales of Landsat data, 136, 178, 179, 182, 187, 192, 195, 200, 205
SAMOS. *See* Satellite Military Observation System
Satellite Military Observation System, 35
Satellites, foreign, xii, 29, 193–195, 204, 210
Saturn, 16
Scanners, 35–36, 48, 49, 55, 67, 70–77, 104, 114, 179, 194, 198, 200, 203, 204, 220n. *See also* Multispectral Scanner
Schmookler, Jacob, 7
Schock, Al, 134
Science, 15–18, 31, 46–47, 53, 54, 81, 92, 96, 97, 117, 123, 125, 127–129, 131, 184, 198, 209, 218n
Science Advisory Committee. *See* President's Science Advisory Committee
Science, Department of, 12
Scientists, 3, 4, 8, 12, 15, 16, 18, 28, 31, 34, 37–39, 40, 42, 49, 51, 53, 54, 56–58, 67, 69, 72, 73, 76, 77, 86, 91, 95, 96, 99, 106, 109, 110 111, 118, 123–131, 134, 137, 142, 143, 150, 152, 155, 157, 164, 172, 174, 177, 185, 189, 191, 198, 199, 202, 208
 as constituency for NASA, 123
 foreign, 177, 180, 185, 189
Seamans, Robert C., Jr., 62, 63
Seasat, 57
Secrecy, industrial, 171, 172, 175
Secrets, military, 61, 186
Security issues, 4, 35, 36, 38, 74, 75, 76, 88, 91
Selection of sensors, 66–79, 208
Senate, 201
Sensors, 4, 11, 27, 28, 33, 35, 37, 38, 42, 45, 47–55, 59, 62–64, 66–79, 80, 82, 84, 85, 87–90, 97, 98–99, 101, 102, 104, 108, 112, 114, 116, 125, 126, 147, 161, 167, 168, 184, 185, 191, 194, 198, 200, 202, 204, 208, 218n, 219n, 220n, 228n
Shapley, Willis H., 61, 151, 204
Shipping, 115, 165, 191
Shoemaker, F. Floyd, 167
Signature extension, 148, 153, 154

Sioux Falls, South Dakota (EROS Data Center), 134, 135, 136, 137, 178
Site selection for data center, 133–136
Skolnikoff, Eugene B., 204
Skylab, 45, 52, 53, 97, 113, 219n
Smith, Constance Babington, 32
Smithsonian Institution, xiii
Snow monitoring, 3, 28, 115, 128, 129, 130. See also Hydrology
Social construction of technology, 4, 8, 107, 121, 132, 212n
Social goals, 4, 36
Soil, 49, 142, 150, 154, 176, 185, 218n
Sos, John Y., 116
South Africa, 192
South America, 59, 188
South Dakota, 134, 135, 136
Soviet Union, 15, 16, 18, 33–35, 146, 150–151, 154, 155, 186, 187, 193, 219n
Space shuttle, 203, 225n, 228n
Space station, 52, 53, 219n
Spectral bands, 47–49, 51, 72, 76–78, 99, 114
Spectral signatures, 47, 48
Spinoff, 163, 164, 177
SPOT (Système Probatoire d'Observation de la Terre), xii, 29, 193–194, 210
Sputnik, 15, 16, 34
Stanford University, 177
State, Department of, 25, 59
State government use of Landsat data, 12, 28, 29, 123, 139, 141, 144, 159–170, 178, 182, 205–206, 208–210
Statistical Reporting Service (Department of Agriculture), 150, 155, 156, 210
Stenseth, David, 134
Storm warnings, 22
Strip mining, 144
Strip planting, 154–155
Sweden, 192
Syncom, 23, 25–27

Tape recorders, 53, 66, 109, 113, 114, 229n
Technological change, 3–14, 116–117, 121–123, 168, 211, 213n
Technology transfer, 6, 8, 12, 29, 35–37, 97, 142, 157, 159–170,
Technology Week, 63–64

Telephone, 23
Telescope, 40
Television, 23, 61, 69, 179, 228n
Television and Infra–Red Observation Satellite (Tiros), 20, 22, 102, 186
Television cameras, 19, 35, 36, 42, 55, 59, 61, 67–71, 193, 194, 215n, 216n, 227n, 228n. *See also* Cameras
Television–type sensors, 59, 67–71, 89. *See also* Return beam vidicon
Telstar, 23, 24
Tennessee Valley Authority, 32
Testing phase of new technology development, 7, 13, 19, 46, 50, 53, 63, 102, 147, 197, 200, 205, 208, 209
Tests of Landsat data use, 147
Texas, 17, 130, 161, 169
Thematic mapper, 78, 194, 198, 200, 203, 204
Thermal sensors, 33, 74, 177
Third World countries, 89, 181, 189. *See also* Developing countries
Tibet, 190
Timeliness of data processing, 108, 110, 113–117, 138, 150, 151, 153, 157
Tiros. *See* Television and Infra–Red Observation Satellite
Training programs, 140–145, 160, 164, 167, 177, 178, 185, 189, 201, 242n
Trust, 12, 59, 135, 152, 159, 163, 166, 168, 169, 174, 181, 188, 209
TRW Corporation (Thompson–Ramo–Wooldridge), 101–104, 117, 232n

U–2, 33, 34, 52, 219n
Udall, Stewart L., 60, 61, 62, 64, 86, 101
Ultraviolet, 54
United Nations, 87, 144, 180, 183, 186–188
United States Geological Survey, 5, 41, 49, 55, 57–65, 68, 73, 74, 77, 86, 90, 98, 133, 134, 142, 144, 160, 184, 185
United States Steel Corporation, 165
Universities, 3, 37, 46, 49–51, 96, 125, 141, 163, 167, 177
User agencies, xi, 5, 6, 56–65, 67, 76, 77, 80, 84, 91, 94–106, 107–113, 125, 126, 129, 132, 150, 162, 182, 184, 197, 199–207, 208, 210

User Communications Task Team, 165
Users, xi, xii, 4, 5, 6, 8, 12–14, 15, 18–19, 22, 23, 27–30, 42, 52, 55, 56–65, 66, 67, 75, 76, 87, 89, 90, 92, 94–106, 107–118, 121–131, 136–145, 145–158, 159–170, 171–182, 183–195, 196–207, 208–211, 219n, 229n, 236n
Usher, Abbot P., 7
USSR. *See* Soviet Union

Vegetation, 32–33, 49, 71, 77, 147, 176, 177, 185, 218n
Video images, 103, 114, 116, 117
Vidicons, 48, 67, 68–78, 133. *See also* Return beam vidicon
Virginia, 17, 142

Wallops Flight Center, 169
Washington (state), 130, 160
Washington, University of, 39
Water, 28, 122, 127–131, 150, 161, 169, 170, 176, 179, 206
Wavelengths, 66, 71, 77
Watkins, Allen, 142, 143
Weather, 18, 20, 21, 54, 110, 142, 146, 151, 153, 154, 157, 171, 206, 213n
Weather Bureau, 6, 19–23, 133
Weather satellites, 3, 6, 15, 18, 19–23, 27, 31, 36, 39, 40, 42, 47, 73, 82, 95, 101, 102, 110, 133, 171, 182, 186, 199, 205, 213n, 215n
Webb, James E., 62, 64, 84, 226n
Weinstein, Oscar, 69, 73
Westinghouse Corporation, 75, 91, 199
Wheat, 47, 49, 109, 146, 150–155
Williamson, David, 74, 198
Willow Run Laboratory, 177. *See also* Institute of Science and Technology
World War I, 31
World War II, 32, 34, 89, 217n
Wyoming, 173